BOB MILLER'S HIGH SCHOOL CALCULUS FOR THE CLUELESS

HIGH SCHOOL CALCULUS

OTHER BOOKS BY BOB MILLER

Bob Miller's Basic Math and Pre-Algebra for the Clueless, Second Edition
Bob Miller's Algebra for the Clueless, Second Edition
Bob Miller's Geometry for the Clueless, Second Edition
Bob Miller's SAT™ Math for the Clueless, Second Edition
Bob Miller's Precalc with Trig for the Clueless, Third Edition
Bob Miller's Calc I for the Clueless, Second Edition
Bob Miller's Calc II for the Clueless, Second Edition
Bob Miller's Calc III for the Clueless

BOB MILLER'S HIGH SCHOOL CALCULUS
FOR THE CLUELESS

HIGH SCHOOL CALCULUS

Honors Calculus, AB and BC Calculus

Robert Miller

Formerly Mathematics Department
City College of New York

New York   Chicago   San Francisc   Lisbon   London   Madrid
Mexico City   Milan   New Delhi   San Juan   Seoul
Singapore   Sydney   Toronto

1 2 3 4 5 6 7 8 9 10 11 12 13 14 15   DOC/DOC   0 9 8 7

ISBN: 978-0-07-148845-7
MHID: 0-07-148845-6

Cataloging-in-Publication Data is on file with the Library of Congress.

To my wonderful wife Marlene.
I dedicate this book and everything else I ever do to you.
I love you very, very much.

ABOUT BOB MILLER...
IN HIS OWN WORDS

After graduating from George W. Hewlett HS, Hewlett, L.I., N.Y., I received my B.S. and M.S. in math from Polytechnic University, Brooklyn, NY. After my first class at Poly, which I taught as a substitute, one student told another upon leaving, "at least we have someone that can teach the stuff." I was forever hooked on teaching. I have taught at C.U.N.Y., Westfield State College and Rutgers. My name is in three editions of *Who's Who Among America's Teachers*. No matter how badly I feel, I always feel great when I teach. I am always delighted when students tell me they hated math before but now they like it and can do it. My main blessing is my family: my fabulous wife Marlene, my wonderful children Sheryl and Glenn, Eric and Wanda, and my delicious grandchildren Kira, Evan, Sean, Sarah, and Ethan. My hobbies are golf, bowling, crossword puzzles, and sudoku. Someday I hope a publisher will allow me to publish the ultimate high school math text and the ultimate calculus text so that all students will understand and benefit from math. It will ensure that our country will remain number one in thinking, in math, and in success. If you can help, please contact me.

To me, teaching math is always a great joy. I hope I can give some of this joy to you.

CONTENTS

ACKNOWLEDGMENTS

I have many people to thank.

I thank my wife, Marlene, who makes life worth living, who is the wind under my wings.

I thank the rest of my family: children, Sheryl and Glenn, Eric and Wanda; grandchildren, Kira, Evan, Sean, Sarah, and Ethan; brother, Jerry; and parents and in-law parents, Cele and Lee, Edith and Siebeth.

I thank the past and present staff at McGraw-Hill: Barbara Gilson, Chuck Wall, John Carleo, John Aliano, David Beckwith, Kimberly Eaton, and Maureen Walker.

I thank Martin Levine for introducing my books to McGraw-Hill.

I thank Dr. Robert Urbanski, Bernice Rothstein, Sy Solomon, and Daryl Davis.

As usual, the last three thanks go to three terrific people: a great friend Gary Pitkofsky, another terrific friend and fellow teacher David Schwinger, and my cousin Keith Robin Ellis.

TO THE STUDENT

This book is written for you: not for your teacher, not for your next door neighbor—not for anyone but you.

However, as much as I hate to admit it, I am not perfect. If you find something that is unclear or a topic that should be added to the book, you can contact me in one of two ways. You can write me c/o McGraw-Hill Professional, Two Penn Plaza, New York, NY 10121-2298. Please enclose a self-addressed, stamped envelope. Be patient; I will answer. You can also contact me at bobmiller@ mathclueless.com. Although my books are not dull, as you will see, my web site, www.mathclueless. com, is. Hopefully by the time you see it, the site will be much better. However, you can visit me at www.mathclueless.com.

I will answer faster than if you write, but, again, please be patient.

Bob Miller

BOB MILLER'S HIGH SCHOOL CALCULUS FOR THE CLUELESS

HIGH SCHOOL CALCULUS

CHAPTER I

THE BEGINNING— LIMITS

INFORMAL DEFINITION

We will begin at the beginning. Calculus starts with the concept of limits. We will examine this first intuitively before we tackle the more difficult theoretical definition.

Let us examine

$$\lim_{x \to a} f(x) = L$$

which is read, "The limit of f(x) as x goes to a is L."

This means that the closer x gets to a, the closer f(x) gets to L. We will leave the word *close* unspecified until later.

EXAMPLE I—

$$\lim_{x \to 3} 2x$$

We will take points near x = 3, smaller than 3, getting closer to 3. We make a small chart showing this.

x	2x
2.5	5
2.9	5.8
2.99	5.98
2.999	5.998

We see that as x approaches 3 from points less than 3, f(x) approaches 6. *Notation*:

$$\lim_{x \to 3^-} f(x) = 6$$

is read, "The limit of f(x) as x goes to 3 from the negative side of 3 (numbers less than 3) is 6." We call this the limit from the left.

If we do the same thing for numbers greater than 3, the chart looks like this:

x	2x
3.2	6.4
3.1	6.2
3.01	6.02
3.001	6.002

The limit from the right

$$\lim_{x \to 3^+} f(x)$$

also equals 6. Since the limit from the left equals the limit from the right, the limit exists and is equal to 6. We write

$$\lim_{x \to 3} f(x) = 6$$

After seeing this example, you might tell me, "Hey, you big dummy!! All you have to do is substitute x = 3 and get the answer!!" Substitution does work sometimes and should always be tried first. However, if limits (and calculus) were so easy, it would not have taken such dynamite mathematicians as Newton and Leibniz to discover calculus.

EXAMPLE 2—

$$\lim_{x \to 4} \frac{x - 4}{x - 4}$$

We first substitute $x = 4$ and get $0/0$, which is indeterminate. We again make a chart.

x	$\dfrac{x - 4}{x - 4}$
4.1	1
4.01	1
4.001	1
3.9	1
3.99	1
3.9999	1

As we get close to 4 from both sides, the answer not only is close to 1 but equals 1. We conclude that the limit as x goes to 4 equals 1.

We get a little better idea of

$$\lim_{x \to a} f(x) = L$$

This means that $f(x)$ is defined at all points very close to a and that the closer x gets to a, the closer $f(x)$ gets to L (if it doesn't already *equal* L).

EXAMPLE 3—

$$\lim_{x \to 2} \frac{x^2 + 3x + 4}{2x + 5}$$

Nothing bad here.

$$\frac{(2)^2 + 3(2) + 4}{2(2) + 5} = \frac{14}{9}$$

EXAMPLE 4—

$$\lim_{x \to 4} \frac{x - 4}{2x + 3} = \frac{0}{11} = 0$$

EXAMPLE 5—

$$\lim_{x \to 2} \frac{x}{x - 2} = \frac{2}{0}$$

which is undefined.

The limit does not exist. The limit must be a number; infinity is not a number.

Let's give one more demonstrated example of what it is to find the limit point by point.

$$\lim_{x \to 2} \frac{x^2 - 4}{5x - 10}$$

First we let x = 2. We find the answer is 0/0. Let's make charts again.

x	$\dfrac{x^2 - 4}{5x - 10}$	x	$\dfrac{x^2 - 4}{5x - 10}$
3	1.0	1	0.6
2.5	0.9	1.5	0.7
2.1	0.82	1.9	0.78
2.01	0.802	1.99	0.798
2.001	0.8002	1.999	0.7998

So

$$\lim_{x \to 2^+} \frac{x^2 - 4}{5x - 10} = 0.8$$

and

$$\lim_{x \to 2^-} \frac{x^2 - 4}{5x - 10} = 0.8$$

Therefore, the limit is 0.8. However, we can't make a chart every time. For Examples 3, 4, and 5, a chart is not necessary. However, Example 6 shows what has to be done sometimes.

Warning: Substitution of a number such as $x = 2$ does not work all the time, especially when you have a function that is defined in pieces, such as that in Example 22 at the end of this chapter. Note that $f(1) = 6$, but

$$\lim_{x \to 1} f(x) = 1$$

Also note that $f(6) = 4$, but the $\lim f(x)$ as x goes to 6 does not exist. So be carrrrreful!!!

EXAMPLE 6—

$$\lim_{x \to 3} \frac{x^2 - 9}{5x - 15}$$

First we substitute $x = 3$ and get 0/0, which is indeterminate. We don't want to make charts all the time. In this case we can factor.

$$\frac{x^2 - 9}{5x - 15} = \frac{(x + 3)(x - 3)}{5(x - 3)}$$

$$\lim_{x \to 3} \frac{(x + 3)}{5} \frac{(x - 3)}{(x - 3)} = \frac{6}{5} \cdot 1 = \frac{6}{5}$$

EXAMPLE 7—

$$\lim_{x \to 0} \frac{x}{|x|}$$

First we substitute x = 0, and we again get 0/0. Making a chart, we get

x	$\dfrac{x}{|x|}$
0.3	1
0.1	1
0.01	1
0.001	1
−0.1	−1
−0.01	−1
−0.0001	−1

The limit from the left is −1, and the limit from the right is 1. Since they are not the same,

$$\lim_{x \to 0} f(x)$$

does not exist. The graph will show that the limit does not exist at x = 0.

EXAMPLE 8—

$$\lim_{x \to 4} \frac{x^{1/2} - 2}{x - 4}$$

There are two ways to do this problem. We can rationalize the numerator, remembering not to multiply out the bottom; or we can factor the bottom into the difference of two squares in a kind of weird way not found in most algebra books today.

METHOD A

$$\frac{(x^{1/2} - 2)}{(x - 4)} \frac{(x^{1/2} + 2)}{(x^{1/2} + 2)} = \frac{(x - 4)}{(x - 4)(x^{1/2} + 2)} = \frac{1}{x^{1/2} + 2}$$

We now take

$$\lim_{x \to 4} \frac{1}{x^{1/2} + 2} = \frac{1}{2 + 2} = \frac{1}{4}$$

METHOD B

$$\frac{x^{1/2} - 2}{x - 4} = \frac{x^{1/2} - 2}{(x^{1/2} - 2)(x^{1/2} + 2)} = \frac{1}{x^{1/2} + 2}$$

which gives the same result.

EXAMPLE 9—

$$\lim_{x \to 3} \frac{5 - 15/x}{x - 3}$$

We will multiply top and bottom by x.

$$\frac{(5 - 15/x)x}{(x - 3)x} = \frac{5x - 15}{(x - 3)x} = \frac{5(x - 3)}{x(x - 3)} = \frac{5}{x}$$

To get the answer, we take the limit.

$$\lim_{x \to 3} \frac{5}{x} = \frac{5}{3}$$

LIMITS AS x GOES TO INFINITY

Although this topic occurs later in your book (and my book), some texts talk about limits at infinity very early on. So I've decided to add this section. If you don't need it now, skip it until later.

We need to know one fact from elementary algebra. The *degree* of a polynomial in one unknown is the highest exponent.

EXAMPLE 10—

$$\lim_{x \to \infty} \frac{4x + x^2}{5x^3 + 1}$$

$$\frac{4x + x^2}{5x^3 + 1} = \frac{4x/x^3 + x^2/x^3}{5x^3/x^3 + 1/x^3} = \frac{4/x^2 + 1/x}{5 + 1/x^3}$$

Divide every term, top and bottom, by x^3, which is x to the highest degree.

If we now take the limit as x goes to infinity, every term goes to 0 except the 5 and we get $(0 + 0)/(5 + 0) = 0$.

IMPORTANT NOTE I

Anytime the degree of the top is less than the degree of the bottom, the limit will be 0. You need not do the work (unless the teacher demands it). You should know that the limit is 0!!!!!!

EXAMPLE I I—

$$\lim_{x \to \infty} \frac{x^2 x^{1/2} + 3x - 1}{x^3 + 5x^2 - 99}$$

If we could talk about a degree of the top, it would be 5/2, or 2½. Since the degree of the bottom is 3, which is more than the top, the limit is 0!

EXAMPLE I 2—

Divide everything by x³.

$$\lim_{x \to \infty} \frac{4x^3 + 7}{x - 3x^3}$$

We get

$$\frac{4x^3/x^3 + 7/x^3}{x/x^3 - 3x^3/x^3} = \frac{4 + 7/x^3}{1/x^2 - 3}$$

If we now let x go to infinity, we find the limit to be 4/(−3) or −4/3.

IMPORTANT NOTE 2

If the degree of the top is the same as the degree of the bottom, the limit is the coefficient of the highest power on the top divided by the coefficient of the highest power on the bottom. Again, you do not actually have to do the division.

Here are two more limits as x goes to infinity.

EXAMPLE I 3—

$$\lim_{x \to \infty} (x^2 + 4)^{1/2} - (x^2 - 1)^{1/2}$$

We get infinity minus infinity. No good!!! What to do?
We rationalize the situation. Seriously, we multiply the
top and bottom by the conjugate.

$$[(x^2 + 4)^{1/2} - (x^2 - 1)^{1/2}] \frac{(x^2 + 4)^{1/2} + (x^2 - 1)^{1/2}}{(x^2 + 4)^{1/2} + (x^2 - 1)^{1/2}}$$

So we get

$$\lim_{x \to \infty} \frac{5}{(x^2 + 4)^{1/2}(x^2 - 1)^{1/2}} = 0$$

EXAMPLE 14, PART A—

$$\lim_{x \to \infty} \frac{(3x^2 + 4x + 5)^{1/2}}{7x}$$

EXAMPLE 14, PART B—

Same example, except x goes to minus infinity.

As x goes to plus or minus infinity, only the highest
power of x counts. So $(3x^2 + 4x + 5)^{1/2}$ is approximately
equal to $3^{1/2} |x|$ for very big and very small values of x.

A. Soooo

$$\lim_{x \to \infty} \frac{3^{1/2} |x|}{7x} = \frac{3^{1/2}}{7}$$

buuuuuuut

B. $\lim\limits_{x \to -\infty} \dfrac{3^{1/2} |x|}{7x} = \dfrac{-3^{1/2}}{7}$!!

PROBLEMS INVOLVING $\lim\limits_{x \to 0} \dfrac{\sin x}{x}$

In proving that the derivative of the sine is the cosine,
which is found in nearly every text, we also prove

$$\lim_{x \to 0} \frac{\sin x}{x} = 1$$

This means if we take the sine of any angle and divide it by *precisely* the same angle—if we now take the limit as we go to 0—the value is 1. For some reason, this topic, which requires almost no writing or calculation, causes a tremendous amount of agony. Hopefully, I can lessen the pain.

FACT 1

$$\lim_{x \to 0} \frac{\sin x}{x} = 1$$

FACT 2

$$\lim_{x \to 0} \sin x = 0$$

FACT 3

$$\lim_{x \to 0} \cos x = 1$$

EXAMPLE 15

$$\lim_{x \to 0} \frac{\sin 3x}{4x}$$

To use Fact 1, since the angle on the top is 3x, the angle on the bottom must also be 3x. So put the 4 on the left and multiply the bottom by the 3 you need. If you multiply the bottom by 3, you must multiply the top by 3 so nothing changes.

Sooooo . . .

$$\frac{\sin 3x}{4x} = \frac{3}{3} \frac{\sin 3x}{4x} = \frac{3}{4} \frac{\sin 3x}{3x}$$

$$\lim_{x \to 0} \frac{3}{4} \frac{\sin 3x}{3x} = \frac{3}{4} \lim_{x \to 0} \frac{\sin 3x}{3x} = \frac{3}{4} (1) = \frac{3}{4}$$

EXAMPLE 16

We use the identity tan x = sin x/cos x.

$$\lim_{x \to 0} \frac{\tan x}{x}$$

$$\frac{\tan x}{x} = \frac{1}{x}\frac{\sin x}{\cos x} = \frac{1}{\cos x}\frac{\sin x}{x}$$

Therefore

$$\lim_{x\to0}\frac{\tan x}{x} = \lim_{x\to0}\frac{1}{\cos x}\frac{\sin x}{x} = \left(\lim_{x\to0}\frac{1}{\cos x}\right)\left(\lim_{x\to0}\frac{\sin x}{x}\right)$$

$$= (1/1) \times 1 = 1$$

How are you doing so far? Let's put in one more example.

EXAMPLE 17—

$$\lim_{x\to0}\frac{\sin^2 x}{x} = \lim_{x\to0}(\sin x)\frac{\sin x}{x} = \left(\lim_{x\to0}\sin x\right)\left(\lim_{x\to0}\frac{\sin x}{x}\right)$$

$$= 0(1) = 0$$

NOTE

In Example 17, if $\sin^2 x$ were in the bottom and x were in the top, then the limit would be 1/0, which would be undefined.

FORMAL DEFINITION

We will now tackle the most difficult part of basic calculus, the theoretical definition of the limit. As previously mentioned, it took two of the finest mathematicians of all times, Newton and Leibniz, to first formalize this topic. It is *not* essential to the rest of basic calculus to understand this definition. It is hoped this explanation will give you some understanding of how really amazing calculus is and how brilliant Newton and Leibniz must have been. Remember this is an approximating process that many times gives exact (or, if not, very, very close) answers. To me this is mind-boggling, terrific, stupendous unbelievable, awesome, cool, and every other great word you can think of.

DEFINITION

$$\lim_{x \to a} f(x) = L$$

If and only if, given $\varepsilon > 0$, there exists $\delta > 0$ such that if $0 < |x - a| < \delta$, then $|f(x) - L| < \varepsilon$.

NOTE

ε = epsilon and δ = delta—two letters of the Greek alphabet.

TRANSLATION 1

Given ε, a small positive number, we can always find δ, another small positive number, such that if x is within a distance δ from a but not exactly at a, then f(x) is within the distance ε from L.

TRANSLATION 2

We will explain this definition by using an incorrect picture. I feel this gives you a much better idea than the correct picture, which we will use next.

Interpret $|x - a|$ as the distance between x and a, but instead of the one-dimensional picture it really is, imagine that there is a circle around the point a of radius δ. Now $|x - a| < \delta$ stands for all x values that are inside this circle. Similarly, imagine a circle of radius ε around L, with $|f(x) - L| < \varepsilon$, the set of all points f(x) that are inside this circle.

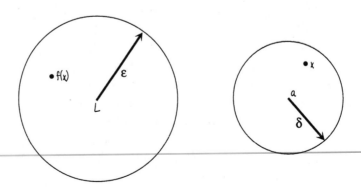

The definition says given $\varepsilon > 0$ (given a circle of radius ε around L), we can find $\delta > 0$ (circle of radius δ around a) such that if $0 < |x - a| < \delta$ (if we take any x inside this circle), then $|f(x) - L| < \varepsilon$ [f(x)] will be inside of the circle of radius ε but not exactly at L.

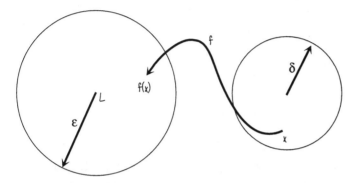

Now take another ε, say ε_2, positive but smaller than ε (a smaller circle around L); there exists another δ, call it δ_2, usually a smaller circle around a, such that if $0 < |x - a| < \delta_2$, then $|f(x) - L| < \varepsilon_2$.

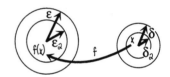

Now take smaller and smaller positive ε's; we can find smaller and smaller δ's. In the limit as the x circle shrinks to a, the f(x) circle shrinks to L. *Read this a number of times!!!*

TRANSLATION 3

Let us see the real picture. $y = f(x)$. $|x - a| < \delta$ means $a - \delta < x < a + \delta$. And $|y - L| < \varepsilon$ means $L - \varepsilon < y < L + \varepsilon$.

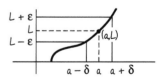

Given $\varepsilon > 0$, if we take any x value such that $0 < |x - a| < \delta$, the interval on the x axis, and find the corresponding $y = f(x)$ value, then this y value must be within ε of L, that is, $|f(x) - L| < \varepsilon$.

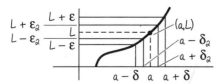

Take a smaller ε_2; we can find δ_2 such that if $0 < |x - a| < \delta_2$, then $|f(x) - L| < \varepsilon_2$. The smaller the ε, the smaller the δ. So $f(x)$ goes to L as x goes to a.

Although this definition is extremely difficult, its application is pretty easy. We need to review six facts, four about absolute value and two about fractions.

1. $|ab| = |a|\,|b|$

2. $\left|\dfrac{a}{b}\right| = \dfrac{|a|}{|b|}, b \neq 0$

3. $|a - b| = |b - a|$

4. $|a + b| \leq |a| + |b|$

5. In comparing two positive fractions, if the bottoms are the same and both numerators and denominators are positive, the larger the top, the larger the fraction. $2/7 < 3/7$.

6. If the tops are the same, the larger the bottom, the smaller the fraction. $3/10 > 3/11$.

Now let's do some problems.

EXAMPLE 18—

Using ε and δ, prove that

$$\lim_{x \to 2} (4x - 3) = 5$$

In the definition

$$\lim_{x \to a} f(x) = L$$

$f(x) = 4x - 3$, $a = 2$, and $L = 5$. Given $\varepsilon > 0$, we must find $\delta > 0$, such that if $0 < |x - 2| < \delta$, then $|(4x - 3) - 5| < \varepsilon$.

$$|(4x - 3) - 5| = |4x - 8| = |4(x - 2)|$$
$$= |4|\,|x - 2| < 4 \cdot \delta = \varepsilon \qquad \delta = \varepsilon/4$$

EXAMPLE 19—

Prove

$$\lim_{x \to 4} (x^2 + 2x) = 24$$

Given $\varepsilon > 0$, we must find $\delta > 0$, such that if $0 < |x - 4| < \delta$, then $|x^2 + 2x - 24| < \varepsilon$.

$$|x^2 + 2x - 24| = |(x + 6)(x - 4)| = |x + 6| \, |x - 4|$$

We must make sure that $|x + 6|$ does not get too big. We must always find δ, no matter how small. We must take a preliminary $\delta = 1$. $|x - 4| < 1$, which means $-1 < x - 4 < 1$ or $3 < x < 5$. In any case $x < 5$.
Sooooo . . .

$$|x + 6| \leq |x| + |6| < 5 + 6 = 11$$

Finishing our problem, we have $|x + 6| \, |x - 4| < 11 \cdot \delta = \varepsilon$. So $\delta = $ minimum $(1, \varepsilon/11)$.

EXAMPLE 20—

Prove

$$\lim_{x \to 5} \frac{2}{x} = \frac{2}{5}$$

$$\left| \frac{2}{x} - \frac{2}{5} \right| = \left| \frac{10 - 2x}{5x} \right| = \frac{|2(5 - x)|}{|5x|} = \frac{|2| \, |5 - x|}{|5| \, |x|}$$

$$= \frac{2|x - 5|}{5|x|}$$

Again take a preliminary $\delta = 1$. $|x - 5| < 1$. So $4 < x < 6$. To make a fraction larger, make the top larger and the bottom smaller. $0 < |x - 5| < \delta$. We substitute δ on the top. Since $x > 4$, we substitute 4 on the bottom.

$$\frac{2\,|x-5|}{5\,|x|} < \frac{2 \cdot \delta}{5 \cdot 4} = \frac{\delta}{10} = \varepsilon \qquad \delta = 10\varepsilon$$

If $\delta =$ minimum $(1, 10\varepsilon)$, then $|2/x - 2/5| < \varepsilon$.

LIMITS AS x GOES TO INFINITY, FORMALLY

Strange as it seems, this might be easier to understand than if x went to a finite number. You might want to look at this before the last section.

Informally, $\lim\limits_{x \to \infty} f(x) = L$ means the larger x gets, the closer f(x) gets to L.

FORMAL DEFINITION

Given an $\varepsilon > 0$, there exists a large positive number N such that whenever $|x| > N$, then $|f(x) - L| < \varepsilon$.

In other words, for every small number ε, we must be able to find a large number N. For numbers bigger than N, f(x) will be very close (ε close) to L.

The picture would look like this:

Show

$$\lim_{x \to \infty} \frac{2x + 5}{x + 1} = 2$$

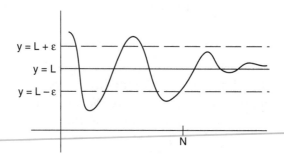

EXAMPLE 21—

Suppose we are given $\varepsilon > 0$. We must find an N such that if $|x| > N$, then $\left|\dfrac{2x + 5}{x + 1} - 2\right| < \varepsilon$. But

$$\left|\frac{2x + 5}{x + 1} - 2\right| = \left|\frac{2x + 5}{x + 1} - \frac{2(x + 1)}{x + 1}\right| = \left|\frac{3}{x + 1}\right| < \varepsilon$$

Since $x > 0$,

$$\left|\frac{3}{x + 1}\right| = \frac{3}{x + 1} < \varepsilon$$

By using inequalities, $\dfrac{3}{x + 1} < \dfrac{\varepsilon}{1}$ is the same as $\dfrac{3}{\varepsilon} < \dfrac{x + 1}{1}$ or $\dfrac{3}{\varepsilon} - 1 < x$. If we let $N = \dfrac{3}{\varepsilon} - 1$, and if $x > N$, then everything previous is true. For any x bigger than this N,

$$\left|\frac{2x + 5}{x + 1} - 2\right| < \varepsilon$$

Note: You may be wondering why $\dfrac{3}{x + 1} < \dfrac{\varepsilon}{1}$ is the same as $\dfrac{3}{\varepsilon} < \dfrac{x + 1}{1}$.

It is true because if $\dfrac{a}{b} < \dfrac{c}{d}$, then $\dfrac{a}{c} < \dfrac{b}{d}$.

A numerical example: Since $\dfrac{2}{7} < \dfrac{3}{4}$, then $\dfrac{2}{3} < \dfrac{7}{4}$.

Again with this part, make sure you know and understand the statements.

Learn the rest at your own pace. Again, despite the fact that you are so good, it might take time, lots of time, to fully understand this part.

THEOREMS ON LIMITS

Throughout this chapter we have been using theorems on limits. We will state all the rules and prove two of them formally.

Let $\lim_{x \to a} f(x) = L$ and $\lim_{x \to a} g(x) = M$. Then

A. $\lim_{x \to a} f(x) + g(x) = L + M$

B. $\lim_{x \to a} f(x) - g(x) = L - M$

C. $\lim_{x \to a} f(x)g(x) = LM$

D. $\lim_{x \to a} f(x)/g(x) = L/M$ if $M \neq 0$

E. $\lim_{x \to a} f(x)^n = L^n$

F. $\lim_{x \to a} \sqrt[n]{f(x)} = L^{1/n}$

Note 1: We took ε/2 for each for the limits so that the limit we were trying to prove has ε in it, as in the technical definition of limits.

Note 2: All of you will get this proof technique with a little work; however, to truly understand that we have proved something, you must understand the technical definition of a limit.

We will prove A and C.

Proof of A: Given an $\varepsilon > 0$, we must find a $\delta > 0$ such that if $0 < |x - a| < \delta$, then $|f(x) + g(x) - (L + M)| < \varepsilon$.

We know $\lim_{x \to a} f(x) = L$. This means given an $\varepsilon/2$, there exists a $\delta_1 > 0$ such that whenever $0 < |x - a| < \delta_1$, then $|f(x) - L| < \varepsilon/2$.

Since $\lim_{x \to a} g(x) = M$, given an $\varepsilon/2$, there is a $\delta_2 > 0$ such that whenever $0 < |x - a| < \delta_2$, then $|g(x) - M| < \varepsilon/2$. Let $\delta = \min(\delta_1, \delta_2)$. For this delta,

$|[f(x) + g(x)] - (L + M)|$

$= |[f(x) - L] + [g(x) - M]|$ (By the triangle inequality)

$\leq |f(x) - L| + |g(x) - M| < \varepsilon/2 + \varepsilon/2 = \varepsilon$

Let's prove C. It is clever and very typical of many techniques used in proving theorems in later math. Again, we have to take very strange-looking ε's in order to have ε at the end.

Proof of C:

$|f(x)g(x) - LM|$

$= |f(x)g(x) - f(x)M + f(x)M - LM|$ (We added zero in a way that helps us.)

$\leq |f(x)(g(x) - M)| + |Mf(x) - LM|$ (Factoring and triangle inequality)

$= |f(x)| |g(x) - M| + |M| |f(x) - L|$ (Property 1 of absolute value, twice)

We need three things to finish the proof:

1. Show that $f(x)$ doesn't get too big.

2. Use the fact that $g(x)$ has a limit.

3. Use the fact that $f(x)$ has a limit.

Here's how we do it:

1. The $\lim_{x \to a} f(x) = L$ means that, in some region near a, $0 < |x - a| < \delta_1$, $|f(x)| < |L| + 1$. [Since the limit of $f(x)$ is L, at some point, $L + 1$ must be bigger than $f(x)$.

2. Since $\lim_{x \to a} g(x) = M$, given an $\varepsilon/2(|L| + 1)$, there exists a δ_2 such that whenever $0 < |x - a| < \delta_2$, then $|g(x) - M| < \varepsilon/2(|L| + 1)$.

3. Again since $\lim_{x \to a} f(x) = L$, given an $\varepsilon/2|M|$, there exists a δ_3 such that whenever $0 < |x - a| < \delta_3$, then $|f(x) - L| < \varepsilon/2|M|$.

   Now let $\delta = $ minimum $(\delta_1, \delta_2, \delta_3)$. So $|f(x)| |g(x) - M| + |M| + |f(x) - L| < (|L| + 1) \varepsilon/2(|L| + 1) + |M|\varepsilon/2|M| = \varepsilon/2 + \varepsilon/2 = \varepsilon$. Whew!

As I said, the techniques will come sooner; the understanding will come later. Let's go on to something simpler, in this case, continuity. Actually simpler things are anything in the rest of this book.

The rest of the δ, ε section is included because you are such good students. However, even the best students

may not be able to cope with these proofs, both understanding them and applying them. Make sure you
understand the statements. Then, over time learn as
much more about them as you can. Fortunately it is
not necessary to understand the proofs in order to
understand the rest of calculus.

CONTINUITY

We finish with a brief discussion of continuity of a
function at a point. Intuitively, continuity at a point
means there is no break in the graph of the function at
the point. Let us define continuity more formally:
f(x) is *continuous* at point a if

1. $\lim\limits_{x \to a} f(x) = L$

2. $f(a) = L$

We will do a longish example to illustrate the definition fully.

EXAMPLE 22—

Let $f(x) = 1$ $x < 1$

　　　　　$= 6$ $x = 1$

　　　　　$= x$ $1 < x < 3$

　　　　　$= 6 - x$ $3 \leq x < 6$

　　　　　$= 4$ $x \leq 6$

We wish to examine the continuity at x = 1, 3, and 6.
  Let's graph this function.

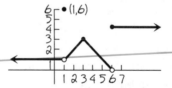

At x = 1, the limit from the left of f(x) = 1 is 1. The limit from the right of f(x) = x is 1. So

$$\lim_{x \to 1} f(x)$$

exists and equals 1. Part 1 of the definition is satisfied. But f(1) = 6. The function is not continuous at 1. (See the jump to y = 6 at x = 1.)

At x = 3, the limits from the left and the right at 3 equal 3. In addition, f(3) = 3. The function is continuous at x = 3. (Notice, no break at x = 3.)

At x = 6, the limit as x goes to 6 from the left is 0. The limit as x goes to 6 from the right is 4. Since the two are different, the limit does not exist. The function is not continuous at 6 (see the jump). We do not have to test the second part of the definition since part 1 fails.

EXAMPLE 23—

Suppose

$$f(x) = \begin{cases} 2x & x < 1 \\ ax + b & 1 \leq x \leq 4 \\ x^3 & x > 4 \end{cases}$$

If a and b are numbers to make f(x) continuous, we must find a and b.

If x is continuous at x = 1, $\lim_{x \to 1^-} 2x = \lim_{x \to 1^+} (ax + b)$. So a + b = 2.

If x is continuous at x = 4, $\lim_{x \to 4^-} (ax + b) = \lim_{x \to 4^+} x^3$. So 4a + b = 64.

Subtracting the first equation from the second gives 3a = 62, a = 62/3, and b = 2 − 62/3 = −56/3.

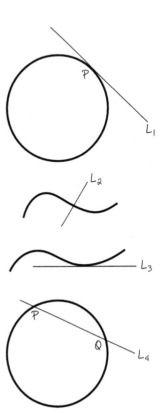

CHAPTER 2
THE BASICS

DERIVATIVES—DEFINITION AND RULES

We would like to study the word *tangent*. In the case of a circle, the line L_1 is tangent to the circle if it hits the circle in one and only one place.

In the case of a general curve, we must be more careful. We wish to exclude lines such as L_2. We wish to include lines such as L_3, even though, if extended, such a line would hit the curve again.

We also need to use the word *secant*. Line L_4 is secant to a circle if it hits the curve in two places.

DEFINITION
Tangent line to a curve at the point P.

    A. Take point P on the curve.

    B. Take point Q_1 on the curve.

    C. Draw PQ_1.

    D. Take $Q_2, Q_3, Q_4, \ldots$, drawing $PQ_2, PQ_3, PQ_4, \ldots$ with Qs approaching P.

E. Do the same thing on the other side of P: Take R_1, R_2, . . . such that R_1 and R_2 are approaching P.

F. If the secant lines on each side approach one line, we say that this line is tangent to the curve at point P.

We would like to develop the idea of tangent algebraically. We will review the development of slope from algebra.

Given points P_1—coordinates (x_1, y_1)—and P_2—coordinates (x_2, y_2)—draw the line segment through P_1 parallel to the x axis and the line segment through P_2 parallel to the y axis, meeting at point Q. Since everything on P_1Q has same y value, the y coordinate of Q is y_1.

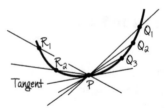

Everything on P_2Q has the same x value. The x value of Q is x_2. The coordinates of Q are (x_2, y_1).

Since everything on P_1Q has the same y value, the length of $P_1Q = x_2 - x_1$. Since everything on P_2Q has the same x value, the length of $P_2Q = y_2 - y_1$. The slope is

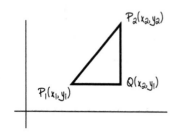

$$m = \frac{\text{change in y}}{\text{change in x}} = \frac{y_2 - y_1}{x_2 - x_1} = \frac{\Delta y}{\Delta x}$$

Δ = delta, another Greek letter

Let's do the same thing for a general function $y = f(x)$.

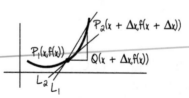

Let point P_1 be the point $(x, y) = (x, f(x))$. A little bit away from x is $x + \Delta x$. (We drew it a lot away; otherwise you could not see it.) The corresponding y value is $f(x + \Delta x)$. So $P_2 = (x + \Delta x, f(x + \Delta x))$. As before, draw a line through P_1 parallel to the x axis and a line through P_2 parallel to the y axis. The lines again meet at Q. As before, Q has the same x value as P_2 and the same y value as P_1. Its coordinates are $(x + \Delta x, f(x))$. Since all y values on P_1Q are the same, the length of $P_1Q = (x + \Delta x) - x = \Delta x$. All x values on P_2Q are the same. The length of $P_2Q = f(x + \Delta x) - f(x)$. The slope of the secant line is

$$L_1 = \frac{\Delta y}{\Delta x} = \frac{P_2Q}{P_1Q} = \frac{f(x + \Delta x) - f(x)}{\Delta x}$$

Now we do as before—let P_2 go to P_1. Algebraically this means to take the limit as Δx goes to 0. We get the slope of the tangent line L_2 at P_1. Our notation will be the slope of the tangent line

$$L_2 = \lim_{\Delta x \to 0} \frac{f(x + \Delta x) - f(x)}{\Delta x}$$

if it exists.

DEFINITION
Suppose $y = f(x)$. The derivative of f(x), at a point x denoted b, $f'(x)$, or dy/dx, is defined as

$$\lim_{\Delta x \to 0} \frac{f(x + \Delta x) - f(x)}{\Delta x}$$

if it exists.

NOTE I
All mathematics originally came from a picture. The idea of derivative came from the slope. Now the definition is independent of the picture.

NOTE 2

If $y = f(t)$ is a distance as a function of time t, then

$$\lim_{\Delta t \to 0} \frac{f(t + \Delta t) - f(t)}{\Delta t} = f'(t)$$

is the velocity v(t).

NOTE 3

In most books today, the notation is

$$\lim_{h \to 0} \frac{f(x + h) - f(x)}{h} = f'(x)$$

The letter h is used instead of Δx, Δy, or Δ, whatever.

NOTE 4

There are alternate definitions for the derivative. However, I like to stick to one in the beginning. It avoids early confusion well, heck. Note 2 about velocity is not enough! Let's do some examples.

EXAMPLE 1

Suppose $y = f(t)$ stands for the distance at some point in time t. Then $f(t + \Delta t)$ stands for your location later, if Δt is positive. (Remember Δt means a change in time.) $y = f(t + \Delta t) - f(t)$ is the distance traveled in time Δt.

$$\frac{\Delta y}{\Delta t} = \frac{f(t + \Delta t) - f(t)}{\Delta t} = \text{average velocity}$$

If we take the limit as Δt goes to 0, that is,

$$\lim_{\Delta t \to 0} \frac{f(t + \Delta t) - f(t)}{\Delta t}$$

then f'(t) is the instantaneous velocity at any time t.

NOTE I

The average velocity is very similar to the rate you learned in elementary algebra. If you took the distance traveled and divided it by the time, you got the rate. The only difference is that in algebra, the average velocity was always the same.

NOTE 2

Even if you drive a car at 30 mph, at any instant you might be going a little faster or slower. This is the instantaneous velocity.

EXAMPLE 2—

Let $f(t) = t^2 + 5t$, with $f(t)$ in feet and t in seconds.

    A. Find the distance traveled between the third and fifth seconds.

    B. Find the average velocity for $3 \leqslant t \leqslant 5$.

    C. Find the instantaneous velocity at $t = 5$.

    A. $\Delta y = f(t + \Delta t) - f(t) = (t + \Delta t)^2 + 5(t + \Delta t)$

$$- (t2 + st)t = 3;\ \Delta t = 5 - 3 = 2$$

$$= t^2 + 2t\,\Delta t + (\Delta t)^2 + 5t + 5\,\Delta t - t^2 - 5t$$

$$= 2t\,\Delta t + (\Delta t)^2 + 5\,\Delta t = 2(3)(2) + (2)^2 + 5(2)$$

$$= 26\ \text{feet}$$

    B. $v_{av} = \Delta y/\Delta t = 26/2 = 13$ feet per second

    C. $v_{inst} = \lim\limits_{\Delta t \to 0} \dfrac{f(t + \Delta t) - f(t)}{\Delta t} = \lim\limits_{\Delta t \to 0} \dfrac{2t\,\Delta t + (\Delta t)^2 + 5\,\Delta t}{\Delta t}$

$$= \lim\limits_{\Delta t \to 0} (2t + \Delta t + 5) = 2t + 5$$

At $t = 5$, $v_{inst} = 2t + 5 = 2(5) + 5 = 15$ feet per second.

NOTE I

Usually Δt is very small when compared to t.

NOTE 2

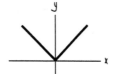

The derivative does not always exist. If $y = |x|$, the derivative does not exist at $x = 0$, since all secant lines on the left have slope -1 and all on the right have slope 1. These lines never approach one line.

In almost all courses, you are asked to do some problems by using the definition of *derivative*. This is really a through exercise in algebra with just a touch of limits. Let us do three examples.

EXAMPLE 3—

Using the definition of *derivative*, find f′(x) if
$f(x) = 3x^2 + 4x - 5$.

$f(x + \Delta x) = 3(x + \Delta x)^2 + 4(x + \Delta x) - 5$

$$f(x + \Delta x) - f(x) = 3(x + \Delta x)^2 + 4(x + \Delta x) - 5 - (3x^2 + 4x - 5)$$

$$= 3x^2 + 6x\,\Delta x + 3(\Delta x)^2 + 4x + 4\,\Delta x - 5$$

$$- 3x^2 - 4x + 5$$

$$= 6x\,\Delta x + 3(\Delta x)^2 + 4\,\Delta x$$

If you have done your algebra correctly, all remaining terms at this point will have at least Δx multiplying them. If there is a fraction, all terms in the top will have at least Δx.

$$\frac{f(x + \Delta x) - f(x)}{\Delta x} = \frac{6x\,\Delta x + 3(\Delta x)^2 + 4\,\Delta x}{\Delta x} = 6x + 3\,\Delta x + 4$$

$$f'(x) = \lim_{\Delta x \to 0} (6x + 3\,\Delta x + 4) = 6x + 4$$

If $f(x) = 3x^2 + 4x - 5$ were a curve, the slope of the tangent line at any point on the curve would be found by multiplying the x value by 6 and adding 4.

EXAMPLE 3 CONTINUED—

Find the slope of the tangent line to the curve
$f(x) = 3x^2 + 4x - 5$ at the point $(3, 34)$.

We don't need to use the 34.

   The slope m $= 6x + 4 = 6(3) + 4 = 22$.

EXAMPLE 3 CONTINUED, CONTINUED—

Find the equation of the line tangent to
$f(x) = 3x^2 + 4x - 5$ at the point $(3, 34)$.

   From algebra, the equation of a line is given by

$$m = \frac{y - y_1}{x - x_1}$$

(point slope). $x_1 = 3$, $y_1 = 34$, and the slope m is
$f'(3) = 22$. The equation of the line is

$$22 = \frac{y - 34}{x - 3}$$

which you can simplify, if forced to.

EXAMPLE 3 LAST CONTINUATION—

Find the equation of the line normal to
$y = 3x^2 + 4x - 5$ at the point $(3, 34)$.

   The word *normal* means to draw the tangent line
at point P and then draw the perpendicular to that
tangent line at point P. Perpendicular slope means
the negative reciprocal. The equation of the normal
line is

$$\frac{-1}{22} = \frac{y - 34}{x - 3}$$

EXAMPLE 4—

$$f(x) = \frac{x}{x + 5}$$

Find f'(x), using the definition of *derivative*.

$$\frac{f(x + \Delta x) - f(x)}{\Delta x} = \frac{\dfrac{x + \Delta x}{x + \Delta x + 5} - \dfrac{x}{x + 5}}{\Delta x}$$

$$= \frac{(x + \Delta x)(x + 5) - x(x + \Delta x + 5)}{(x + \Delta x + 5)(x + 5)(\Delta x)}$$

$$= \frac{5\,\Delta x}{(x + \Delta x + 5)(x + 5)(\Delta x)}$$

$$= \frac{5}{(x + \Delta x + 5)(x + 5)}$$

We must multiply out the top, but we do not multiply the bottom.

$$f'(x) = \lim_{\Delta x \to 0} \frac{5}{(x + \Delta x + 5)(x + 5)} = \frac{5}{(x + 5)(x + 5)} = \frac{5}{(x + 5)^2}$$

EXAMPLE 5—

Find g'(x) by using the definition of *derivative* if

$$g(x) = \sqrt{x}.$$

Rationalize the numerator.

$$\frac{g(x + \Delta x) - g(x)}{\Delta x} = \frac{\sqrt{x + \Delta x} - \sqrt{x}}{\Delta x}$$

$$= \frac{(\sqrt{x + \Delta x} - \sqrt{x})}{\Delta x} \frac{(\sqrt{x + \Delta x} + \sqrt{x})}{(\sqrt{x + \Delta x} + \sqrt{x})}$$

$$= \frac{(x + \Delta x) - x}{(\Delta x)(\sqrt{x + \Delta x} + \sqrt{x})}$$

$$= \frac{\Delta x}{(\Delta x)(\sqrt{x + \Delta x} + \sqrt{x})}$$

$$= \frac{1}{\sqrt{x + \Delta x} + \sqrt{x}}$$

$$g'(x) = \lim_{\Delta x \to 0} \frac{1}{\sqrt{x + \Delta x} + \sqrt{x}}$$

$$= \frac{1}{\sqrt{x} + \sqrt{x}} = \frac{1}{2\sqrt{x}}$$

We can't keep using the definition of derivative. If we had a complicated function, it would take forever. We will list the rules, interpret them, and give examples. Proofs are found in most calculus books.

RULE 1
If $f(x) = c$, then $f'(x) = 0$. The derivative of a constant is 0.

RULE 2
If $f(x) = x$, then $f'(x) = 1$.

RULE 3
If $f(x) = x^n$, then $f'(x) = nx^{n-1}$. Bring down the exponent and subtract 1 to get the new exponent.

Question for you to think about: What does 2^p mean?

EXAMPLE 6—

A. If $y = x^6$,   then $y' = 6x^5$.

B. If $y = x^{-7}$,   then $y' = -7x^{-8}$.

C. If $y = x^{5/3}$,   then $y' = \dfrac{5}{3}x^{2/3}$.

D. If $y = x^{\pi}$,   then $y' = \pi x^{\pi-1}$.

We know $2^{3.14}$ means the hundredth root of 2 raised to the 314th power.

While we wouldn't want to do it, we know how to do it.

Let us prove the power rule for positive integers.
Suppose $f(x) = x^n$. When we prove a derivative rule, we go back to the definition of derivative.

But what does $2^{3.141452653...}$ mean?

$$\text{Proof: } \frac{f(x + h) - f(x)}{h} = \frac{(x + h)^n - x^n}{h}$$

The hint is that you need know nothing more than what has occurred so far in the book.

Expanding by using the binomial theorem, we get

The answer is at the end of the chapter.

$$\frac{x^n + \dfrac{n}{1}x^{n-1}h + \dfrac{n(n-1)}{1(2)}x^{n-2}h^2 + \text{terms multiplied by } h^3 \text{ or more} - x^n}{h}$$

The first and last terms cancel. Dividing all remaining terms by h, we get

$$\lim_{h \to 0} \left[nx^{n-1} + \frac{n(n-1)}{1(2)} x^{n-2}h + \text{terms multiplied by } h^2 \text{ or more} \right]$$

In taking the limit, all terms go to 0 except the first.

We have shown for positive n, if $y = x^n$, its derivative is nx^{n-1}. Let's do the same for negative integers. I only do this for good students. Suppose $f(x) = x^n$, with n a negative integer; then $f'(x) = nx^{n-1}$.

If n is negative, let $n = -m$, where m is a positive integer.

$$f(x) = x^n = x^{-m} = \frac{1}{x^m}$$

$$\frac{f(x + h) - f(x)}{h} = \frac{\dfrac{1}{(x + h)^m} - \dfrac{1}{x^m}}{h}$$

Using algebra found in *Precalc*, we get

$$\frac{f(x + h) - f(x)}{h} = \frac{x^m - (x + h)^m}{x^m(x + h)^m h}$$

Again expanding by using the binomial theorem, we get the right-hand side

$$= \frac{x^m - x^m - \dfrac{m}{1} x^{m-1}h - \dfrac{m(m-1)}{1(2)} x^{m-2}h^2 - \text{terms multiplied by } h^3 \text{ or higher}}{x^m(x + h)^m h}$$

Again the first two terms in the numerator cancel. Divide all remaining terms by h. We get

$$\lim_{h \to 0} \frac{\left[-mx^{m-1} - \dfrac{m(m-1)}{1(2)} x^{m-2}h - \text{terms multiplied by } h^2 \text{ or higher} \right]}{x^m(x + h)^m}$$

As before, all the terms in the numerator drop except for the first one. $(x + h)^m$ goes to x^m. The fraction becomes

$$\frac{-mx^{m-1}}{x^m x^m} = -mx^{m-1-2m} = -mx^{-m-1}$$

But $-m = n$!!!! So the derivative is nx^{n-1} for negative integers. A cute proof!!!! Rational and real numbers will wait for later.

RULE 4

If $f(x) = cg(x)$, then $f'(x) = cg'(x)$. If we have a constant c multiplying a function, we leave c alone and take the derivative of only the function.

RULE 5

If $f(x) = g(x) \pm h(x)$, then $f'(x) = g'(x) \pm h'(x)$.

In a book, t, u, v, w, x, y, and z are usually variables; a, b, c, and k (since mathematicians can't spell) are used as constants.

EXAMPLE 7—

Find y' if

$$y = 3x^7 + 7x^4 + 2x + 3$$

$$y' = 21x^6 + 28x^3 + 2(+0)$$

EXAMPLE 8—

Find y' if

$$y = 3x^4 + \frac{a}{x^6} + \frac{1}{bx^7} + \sqrt[4]{x^9} + \frac{1}{\sqrt[11]{x^5}} + \pi^7$$

$$y = 3x^4 + ax^{-6} + \frac{1}{b}x^{-7} + x^{9/4} + x^{-5/11} + \pi^7$$

$$y' = 12x^3 - 6ax^{-7} + \frac{1}{b}(-7x^{-8}) + \left(\frac{9}{4}\right)x^{5/4} + \left(-\frac{5}{11}\right)x^{-16/11} + 0$$

$$y' = 12x^3 - \frac{6a}{x^7} - \frac{7}{bx^8} + \frac{9x^{5/4}}{4} - \frac{5}{11x^{16/11}}$$

a and 1/b are constants.
9/4 −1 = 5/4.

−5/11 −1 = −16/11.
Derivative of messy constants is still 0.

Most calculus books give the derivative of the six trigonometric functions near the beginning. So will we.

RULE 6

A. If $y = \sin x$, $y' = \cos x$.

B. If $y = \cos x$, $y' = -\sin x$.

C. If $y = \tan x$, $y' = \sec^2 x$.

D. If $y = \cot x$, $y' = -\csc^2 x$.

E. If $y = \sec x$, $y' = \tan x \sec x$.

F. If $y = \csc x$, $y' = -\cot x \csc x$.

Once you prove A, B is proved by using the identities $\cos x = \sin(\pi/2 - x)$, $\sin x = \cos(\pi/2 - x)$, and rule 9.

Once A and B are proved, C through F are proved by writing those four in terms of $\sin x$ and $\cos x$, by using rule 8 and other basic trig identities.

However, I always like to prove that the derivative of the sine is the cosine. It's not the proof that is important but the parts of the proof. Some you know; others you need to review; and still others are new. All are important.

Let's show that if $f(x) = \sin x$, then $f'(x) = \cos x$. As usual, we must use the definition of derivative.

$$\lim_{h \to 0} \frac{f(x + h) - f(x)}{h} = \lim_{h \to 0} \frac{\sin(x + h) - \sin x}{h}$$

$$\sin(A + B) = \sin A \cos B + \cos A \sin B$$

$$= \lim_{h \to 0} \frac{\sin x \cos h + \cos x \sin h - \sin x}{h}$$

Factoring out $\sin x$ from the first and third terms, and splitting by using short division, we get

$$\lim_{h \to 0} \frac{\sin x \, (\cos h - 1)}{h} + \lim_{h \to 0} \cos x \, \frac{\sin h}{h}$$

If we could show

$$\lim_{h \to 0} \frac{\cos h - 1}{h} = 0 \text{ and } \lim_{h \to 0} \frac{\sin h}{h} = 1,$$

our theorem would be proved.

LEMMA I

$$\lim_{h \to 0} \frac{\sin h}{h} = 1$$

LEMMA 2

$$\lim_{h \to 0} \frac{\cos h - 1}{h} = 0$$

Temporarily we will assume lemma 1 is true.

$$\frac{\cos h - 1}{h} = \frac{(\cos h - 1)(\cos h + 1)}{h(\cos h + 1)} = \frac{\cos^2 h - 1}{h(\cos h + 1)}$$

$$= \frac{-\sin^2 h}{h(\cos h + 1)} = \frac{-\sin h}{h} \times \frac{\sin h}{\cos h + 1}$$

So

$$\lim_{h \to 0} \frac{\cos h - 1}{h} = \lim_{h \to 0} \frac{-\sin h}{h} \times \lim_{h \to 0} \frac{\sin h}{\cos h + 1}$$

Now $\lim_{h \to 0} \sin h = 0$, $\lim_{h \to 0} \cos h = 1$, and, by lemma 1,

$\lim_{h \to 0} \frac{\sin h}{h} = 1$. So

$$\lim_{h \to 0} \frac{\cos h - 1}{h} = \lim_{h \to 0} \frac{-\sin h}{h} \times \lim_{h \to 0} \frac{\sin h}{\cos h + 1}$$

$$= (-1)\frac{0}{1 + 1} = (-1)(0) = 0$$

First let's review some terminology.

A theorem is a proven law.

A corollary is a small theorem that is the result of a bigger theorem.

A smaller theorem used to prove a major theorem is called a lemma.

In this case we have two smaller theorems, a dilemma and a dilemma. (A joke.)

If lemma 1 is true, then lemma 2 is true. Now let's show lemma 1 is true.

We now really need the unit circle (probably it made your life miserable in trig). To review, the unit circle is the circle, center at the origin, with radius 1: $x^2 + y^2 = 1$. The center of the circle is O. Point A is (1, 0). Point B = (x, y).

The angle of the sector OAB is h.

$\cos h = x/r = x/1 = x$ and $\sin h = y/r = y/1 = y$ since r = 1 on the unit circle.

So any point (x, y) on this circle is given by (cos h, sin h).

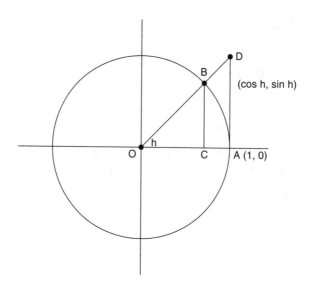

Look at the picture at the area of $\triangle OBC \leqslant$ sector OAB $\leqslant \triangle OAD$.

Let's review the area of a sector. If h = 63°, the area of the sector = $63/360 \times \pi r^2$.

In general the area would be $h/360 \times \pi r^2$.

Angle h is in radians. The area is $h/2\pi \times \pi r^2$. Since r = 1, the area of this sector is h/2.

Now the area of $\triangle OAC \leqslant$ sector $OAB \leqslant \triangle OAD$ or

$$\left(\frac{1}{2}\right)(\cos h)(\sin h) \leqslant \left(\frac{1}{2}\right)h \leqslant \left(\frac{1}{2}\right)\overline{AD}$$

To get $\overline{AD}$, we note that the smaller and larger triangles are similar; we get

$$\frac{y}{x} = \frac{\sin h}{\cos h} = \frac{\overline{AD}}{\overline{OA}} = \frac{\overline{AD}}{1} = \overline{AD}$$

Replace $\overline{AD}$ by $\dfrac{\sin h}{\cos h}$, multiply the whole inequality by 2, and divide by $\sin h$.

We get

$$\cos h \leqslant \frac{h}{\sin h} \leqslant \frac{1}{\cos h}$$

Taking limits, we get

$$\lim_{h \to 0} \cos h \leqslant \lim_{h \to 0} \frac{h}{\sin h} \leqslant \lim_{h \to 0} \frac{1}{\cos h} \quad \text{or} \quad 1 \leqslant \lim_{h \to 0} \frac{h}{\sin h} \leqslant \frac{1}{1}$$

We also notice that the only way that $u \leqslant v$ and at the same time $v \leqslant u$ is when $u = v$. So $\lim\limits_{h \to 0} \dfrac{h}{\sin h} = 1$.

Therefore $\lim\limits_{h \to 0} \dfrac{\sin h}{h}$ since it is the reciprocal.

Lemma 1 is true. Lemma 2 is true. Finally now the derivative of the sine is the cosine. Whew!!

NOTE I

As you can see, there are many facts in this theorem to know.

NOTE 2

After you finish with all the derivative rules, you should show the others are true.

NOTE 3

We already talked about lemma 1 in Chapter 1. This just verifies the fact.

NOTE 4

You sharp-eyed readers might have noticed we proved the statement for positive angles.

$$\text{If } x > 0, \frac{\sin(-x)}{-x} = \frac{-\sin x}{-x} = \frac{\sin x}{x}$$

would prove lemma 1 for negative angles.

Let's go on to the next rule.

RULE 7

The product rule. If $y = f(x)g(x)$, then $y' = f(x)g(x)' + g(x)f(x)'$. The product rule says the first function multiplied by the derivative of the second and added to the second function multiplied by the derivative of the first.

EXAMPLE 9—

Find y' if

$$y = (x^2 + 3x + 1)(5x + 2)$$
$$\quad\quad \text{(first)} \quad\quad \text{(second)}$$
$$y' = (x^2 + 3x + 1) \quad (5) \quad + (5x + 2) \quad (2x + 3)$$
$$\quad\quad \text{(first)} \quad\quad \text{(second)}' \quad \text{(second)} \quad \text{(first)}'$$

Multiplying and combining like terms, we find $y' = 15x^2 + 34x + 11$.

  We could, of course, multiply this example out.

$$y = 5x^3 + 17x^2 + 11x + 2$$

Then $y' = 15x^2 + 34x + 11$, as before.

However, later on, the examples will be much longer or even impossible to multiply out. It is a blessing that we have the product rule and the next two rules.

Let's prove the product rule.

Again we must go back to the definition of derivative.

$$\lim_{h \to 0} \frac{f(x + h)g(x + h) - f(x)g(x)}{h}$$

Using a previous trick of adding zero, we get

$$= \frac{f(x + h)g(x + h) - f(x + h)g(x) + f(x + h)g(x) - f(x)g(x)}{h}$$

Splitting the first two terms from the last two (short algebraic division), and taking a common factor out of each numerator, we get

$$= \lim_{h \to 0} f(x + h) \left[\frac{g(x + h) - g(x)}{h} \right] + \lim_{h \to 0} g(x) \left[\frac{f(x + h) - f(x)}{h} \right]$$

Using the limit theorems we learned before, we have to find four limits. $\lim_{h \to 0} f(x + h)$ is just $f(x)$. Let's skip the second one for a moment. $\lim_{h \to 0} g(x)$ is just $g(x)$.

$$\lim_{h \to 0} \frac{f(x + h) - f(x)}{h} = f'(x)!!!!! \text{ So } \lim_{h \to 0} \frac{g(x + h) - g(x)}{h} = g'(x)!$$

Substituting, we get the product rule $f(x)g'(x) + g(x) f'(x)$.

RULE 8
The quotient rule.

If $y = \dfrac{f(x)}{g(x)}$, then $y' = \dfrac{g(x)f'(x) - f(x)g'(x)}{[g(x)]^2}$.

The quotient rule says the bottom times the derivative of the top minus the top times the derivative of the bottom, all divided by the bottom squared.

EXAMPLE 10—

Find y' if $y = \dfrac{x^2}{x^3 + 2x + 1}$.

(bottom)(top)′
$\dfrac{\quad - (top)(bottom)'}{\text{bottom squared}}$

$$y' = \frac{(x^3 + 2x + 1)(2x) - x^2(3x^2 + 2)}{(x^3 + 2x + 1)^2}$$

When you are simplifying, do *not* multiply out the bottom. Multiply and simplify only the top. You may simplify the top by factoring, as we will do in other problems.

Simplified,

$$y' = \frac{-x^4 + 2x^2 + 2x}{(x^3 + 2x + 1)^2}$$

RULE 9

The chain rule. Suppose we have a composite function $y = f(u)$, $u = u(x)$. Then

$$\frac{dy}{dx} = \frac{dy}{du} \cdot \frac{du}{dx}$$

EXAMPLE 11—

Let $f(x) = (x^2 + 1)^{100}$.

One way is to multiply this out. We dismiss this on the grounds of sanity.

We let $u = x^2 + 1$. Then $y = f(u) = u^{100}$.

$$\frac{du}{dx} = 2x \qquad \frac{dy}{du} = 100u^{99}$$

Then

$$\frac{dy}{dx} = \frac{dy}{du}\frac{du}{dx} = (100u^{99})(2x) = 100(x^2 + 1)^{99} \cdot 2x$$

$$= 200x\,(x^2 + 1)^{99}$$

We don't want to write u each time. We will imagine what u is and use the chain rule. Try it. It only takes a little practice.

EXAMPLE 12—

Find y' if $y = (x^3 + 7x^2 + 1)^{4/3}$.
Imagine $u = x^3 + 7x^2 + 1$.

$$\frac{du}{dx} = 3x^2 + 14x$$

$$y' = \frac{4}{3}\,(x^3 + 7x^2 + 1)^{1/3}\,(3x^2 + 14x)$$

EXAMPLE 13—

$$y = \tan\,(x^4 + 3x - 11)$$

Imagine $u = x^4 + 3x - 11$.

$$y' = \sec^2\,(x^4 + 3x - 11) \cdot (4x^3 + 3)$$

EXAMPLE 14—

Find y' if $y = \sin^6\,(x^4 + 3x)$.

This is a double composite: a function of a function of a function. We use the chain rule twice.
Let the crazy angle = $v = x^4 + 3x$. So $dv/dx = 4x^3 + 3$.
Let $u = \sin\,(x^4 + 3x) = \sin v$. So $du/dv = \cos v$.
So $y = u^6$ and $dy/du = 6u^5$. Thus

$$\frac{dy}{dx} = \frac{dy}{du} \times \frac{du}{dv} \times \frac{dv}{dx}$$

$$= 6u^5 \times \cos v \times 4x^3 + 3$$

$$= [6\sin^5\,(x^4 + 3x)][\cos(x^4 + 3x)]\,(4x^3 + 3)$$

power rule— leave trig function and crazy angle alone	derivative of trig function— crazy angle stays	derivative of crazy angle

NOTE

This is *not* the product rule.

Sometimes you will have other combinations of the rules. After a short while, you will find the rules relatively easy. However, the algebra does require practice.

EXAMPLE 15—

Find y' if $y = (x^2 + 1)^8 (6x + 7)^5$.

This problem involves the product rule. But in each derivative, we will have to use the chain rule.

$$y' = \underbrace{(x^2 + 1)^8}_{\text{(first)}} \underbrace{5(6x + 7)^4(6)}_{\text{(second)}'} + \underbrace{(6x + 7)^5}_{\text{(second)}} \underbrace{8(x^2 + 1)^7(2x)}_{\text{(first)}'}$$

The calculus is now finished. We must simplify by factoring. There are two terms, each underlined. From each we must take out the largest common factor. The largest number that can be factored out is 2. No x can be factored out. The lowest powers of $x^2 + 1$ and $6x + 7$ can be factored out. We take out $(x^2 + 1)^7$ and $(6x + 7)^4$.

$$y' = 2(x^2 + 1)^7 (6x + 7)^4 [15(x^2 + 1) + 8x(6x + 7)]$$

$$= 2(x^2 + 1)^7 (6x + 7)^4 (63x^2 + 56x + 15)$$

Let us try one more, using the quotient rule and chain rule.

EXAMPLE 16—

Find y' if $y = \dfrac{x^3}{(x^2 + 1)^4}$.

$$y' = \frac{(x^2 + 1)^4\, 3x^2 - x^3[4(x^2 + 1)^3(2x)]}{[(x^2 + 1)^4]^2}$$

$$= \frac{x^2(x^2 + 1)^3[3(x^2 + 1) - 8x^2]}{(x^2 + 1)^8}$$

$$= \frac{x^2(-5x^2 + 3)}{(x^2 + 1)^5}$$

When *not* to use the product or quotient rule:

EXAMPLE A—

$y = 5(x^2 - 4)^{10}$

$y' = 5[10(x^2 - 4)^9\,(2x)] = 100x\,(x^2 - 4)^9$

Do not use product rule since 5 is a constant. Only the chain rule is necessary.

EXAMPLE B—

$y = 7/x^5$

$y' = 7(-5)x^{-6} = -35/x^6$

Do not use quotient rule. Rewrite the example as $y = 7x^{-5}$ and simply use the power rule.

EXAMPLE C—

$y = b/(x^2 + 5)^8$

Rewrite as $y = b(x^2 + 5)^{-8}$.

$y' = b(-8)(x^2 + 5)^{-9}\,(2x) = -16bx/(x^2 + 5)^9$

IMPLICIT DIFFERENTIATION

Suppose we are given $y^3 + x^4y^7 + x^3 = 9$. It would be difficult, maybe impossible, to solve for y. However, there is a theorem called the implicit function theorem that gives conditions that will show that $y = f(x)$ exists even if we can never find y. Moreover, it will allow us to find dy/dx even if we can never find y. Pretty amazing, isn't it?!!!!!

Let $f(y) = y^n$ where $y = y(x)$. Using the chain rule and power rule, we get

$$\frac{df}{dx} = \frac{df(y)}{dy}\frac{dy(x)}{dx} = ny^{n-1}\frac{dy}{dx}$$

EXAMPLE 17—

Find dy/dx if $y^3 + x^4y^7 + x^3 = 9$.

1. We will differentiate straight across implicitly.

2. We will differentiate the first term implicitly, the second term implicitly and with the product rule, and the rest the old way.

3. We will solve for dy/dx by using an algebraic trick that can save up to five algebraic steps. With a little practice, you can save a lot of time!!!!

Let's do the problem.

$$3y^2\frac{dy}{dx} + \left[x^4(7y^6)\frac{dy}{dx} + y^7(4x^3)\right] + 3x^2 = 0$$

| implicit (n = 3) | implicit (n = 7) and product rule | power rule | derivative of constant |

We now solve of dy/dx. Once we take the derivative, it becomes an elementary algebra equation in which we solve for dy/dx.

1. All the terms without dy/dx go to the other (right) side and change signs.

2. All terms without dy/dx on the other side stay there; there is no sign change.

3. All terms with dy/dx on the right go to the left and change signs.

4. All terms with dy/dx on the left stay there; there is no sign change.

5. Factor out dy/dx from all terms on the left; this coefficient is divided on both sides. Therefore it goes to the bottom of the fraction on the right.

6. Rearrange all terms so that the number is first and each letter occurs alphabetically.

It really is easy with a little practice. Using this method, we find

$$\frac{dy}{dx} = \frac{-4x^3y^7 - 3x^2}{3y^2 + 7x^4y^6}$$

EXAMPLE 17 CONTINUED—

Maybe you think this is too complicated. Let's do another with much simpler coefficients after taking the derivative: $Ay' + B - Cy' - D = 0$.

B and D have no y' and must go to the other side and flip signs. The $A - C$ is factored out from the y' and goes to the bottom of the answer. So $y' = (D - B)/(A - C)$.

Still don't believe? Let's do it step by step.

$$Ay' + B - Cy' - D = 0$$

$$\underline{\quad\quad -B \quad\quad\quad + D = D - B}$$

$$Ay' \quad\quad - Cy' \quad\quad = D - B$$

$$\frac{(A - C)y'}{A - C} = \frac{D - B}{A - C}$$

So

$$y' = \frac{D - B}{A - C}$$

If you look closely, this really is Example 15, the algebraic part.

EXAMPLE 17 LAST CONTINUATION—

Suppose we are given the same equation, but are asked to find dy/dx at the point $(-2, 1)$.

The first step is the same.

$$3y^2 \frac{dy}{dx} + x^4 \left(7y^6 \frac{dy}{dx} \right) + y^7(4x^3) + 3x^2 = 0$$

But instead of doing all the rest of the work, we substitute $x = -2$ and $y = 1$!

$$3 \frac{dy}{dx} + 112 \frac{dy}{dx} - 32 + 12 = 0$$

So $dy/dx = 20/115 = 4/23$. We do as little work as possible. Mathematicians are very lazy by nature.

We now can show that if $y = x^n$, for n rational, then the derivative is nx^{n-1}.

Let $n = p/q$, for p and q integers, $q \neq 0$.

$$y = x^n = x^{p/q}$$

Taking the qth power of both sides, we get $y^q = x^p$.

Differentiating implicitly, we get $qy^{q-1} \dfrac{dy}{dx} = px^{p-1}$.

Solving, we get

$$\frac{dy}{dx} = \frac{px^{p-1}}{qy^{q-1}} = \frac{p}{q} \frac{x^{p-1}}{x^{p(q-1)/q}} = \frac{p}{q} \frac{x^{p-1}}{x^{p-p/q}} = \frac{p}{q} x^{p-1-p+p/q} = \frac{p}{q} x^{p/q-1} = nx^{n-1}$$

since $n = p/q$. It is quite an algebraic exercise.

Notations

We know the notations for the first derivative. Suppose we want to take more derivatives . . .

$$y' \quad = f'(x) \quad = \frac{dy}{dx} \quad = \text{first derivative}$$

$$y'' \quad = f''(x) \quad = \frac{d^2y}{dx^2} \quad = \text{second derivative}$$

$$y^{(27)} \quad = f^{(27)}(x) \quad = \frac{d^{27}y}{dx^{27}} \quad = \text{twenty-seventh derivative}$$

$$y^{(n)} \quad = f^{(n)}(x) \quad = \frac{d^ny}{dx^n} \quad = \text{nth derivative}$$

EXAMPLE 18—

Find $f^{(30)}(x)$ if $f(x) = \sin(5x)$.

First we notice that the derivatives of the sine function are cyclic (a cycle), and the cycle is 4.

$$f(x) = \sin x, \quad f'(x) = \cos x, \quad f''(x) = -\sin x,$$

$$f'''(x) = -\cos x, \quad f''''(x) = \sin x$$

So 4 into 30 has a remainder of 2 (same as the second derivative). So we have $-\sin 5x$.

Also each time you take the derivative, we get a 5 coming down.

Thirty times means we have 5^{30}. $f^{(30)}(x) = -5^{30} \sin(5x)$.

EXAMPLE 19A—

Let us take three derivatives implicitly of $x^2 - y^2 = 9$. (Something nice usually happens in the even derivatives, in this case the second.)

$$x^2 - y^2 - 9 = 0 \qquad 2x - 2y\frac{dy}{dx} = 0 \qquad \frac{dy}{dx} = \frac{x}{y}$$

Quotient rule:

$$\frac{d^2y}{dx^2} = \frac{y(1) - x(dy/dx)}{y^2}$$

Multiply every term top and bottom by y.

$$= \frac{y - x(x/y)}{y^2}$$

From original equation $x^2 - y^2 = 9$; therefore $y^2 - x^2 = -9$.

$$= \frac{y^2 - x^2}{y^3}$$

(Nice simplification— easy to take next derivative.)

$$= \frac{-9}{y^3}$$

$$\frac{d^3y}{dx^3} = -9(-3)y^{-4}\frac{dy}{dx}$$

$$= 27y^{-4}\left(\frac{x}{y}\right)$$

$$= \frac{27x}{y^5}$$

EXAMPLE 19B—

If $x = 5$ and $y > 0$, find the value of $y'' - y'''$.

This would be a continuation of Example 18 A. Substituting $x = 5$ into $x^2 - y^2 = 9$, we get $y = 4$.

Since $y'' = -\dfrac{9}{y^3}$, $y'' = \dfrac{-9}{4^3} = \dfrac{-9}{64}$.

Since $y''' = \dfrac{27x}{y^5} = \dfrac{27(5)}{4^5} = \dfrac{135}{1024}$,

$f''(x) - f'''(x)$, when $x = 5$,

$$= \frac{-9}{64} - \frac{135}{1024} = \frac{-9 \times 16}{64 \times 16} - \frac{135}{1024} = -\frac{279}{1024}.$$

EXAMPLE 20

$\sin (x^2y^3) - x^4y^5 = 0$

$\cos (x^2y^3) \, [(x^2 3y^2 \dfrac{dy}{dx} + y^3 2x)] - x^4 5y^4 \dfrac{dy}{dx} - y^5 4x^3 = 0$

The derivative of the sine of a crazy angle is the cosine of that crazy angle times the derivative of the crazy angle—implicitly and with the product rule.

The product rule; the minus sign in front means both parts of the product are negative.

Solving for dy/dx in one step, we get

$$\dfrac{dy}{dx} = \dfrac{4x^3y^5 - 2xy^3 \cos(x^2y^3)}{3x^2y^2 \cos(x^2y^3) - 5x^4y^4}$$

The only difference in the algebra is to multiply out the terms in parentheses in your head (the cosine multiplies each term). The example is then virtually the same as Example 15.

Now seems to be a fine time to insert problems you are likely to see. They are written in roughly increasing order of difficulty.

EXAMPLE 21

A. Find the equation of the line tangent to $y = x^2 + 4x + 7$ at the point (1, 12).

B. Find the equation of the line tangent to $y = x^2 + 4x + 7$ if $x = 2$.

C. Find the equation of the lines tangent to $y = x^2 + 4x + 7$ when $y = 12$.

D. Find the equation of the line tangent to $y = x^2 + 4x + 7$ that is perpendicular to the line $5x + 10y = 11$.

E. Find the equation of the two lines passing through point (5, 9), tangent to $y = x^2$.

F. Find the equations of the two lines that are tangent to both $y = x^2$ and $y = -x^2 + 6x - 5$.

Here are the solutions. All answers are written in point-slope form, the easiest way.

Your teacher may want a different form.

A. $m = y' = 2x + 4$. Since $x = 1$, $m = 6$. The line is $6 = \dfrac{y - 12}{x - 1}$.

B. $m = y' = 2x + 4 = 2(2) + 4 = 8$. $y = (2)^2 + 4(2) + 7 = 19$. The line is $8 = \dfrac{y - 19}{x - 2}$.

C. $m = 2x + 4$. Since $x^2 + 4x + 7 = 12$, $x^2 + 4x = 5 = 0$ or $(x + 5)(x - 1) = 0$.

So $x = 1$ and $x = -5$. The points are $(1, 12)$ and $(-5, 12)$.

For the point $(1, 12)$, $m = 2(1) + 4 = 6$ and the equation of the line is $6 = \dfrac{y - 12}{x - 1}$.

For the point $(-5, 12)$, $m = 2(-5) + 6 = -4$ and the equation is $-4 = \dfrac{y - 12}{x + 5}$.

Here's the picture: See, there are two lines where $y = 12$.

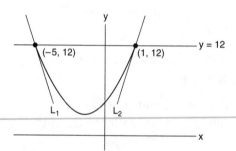

D. For the line $5x + 10y = 11$. Solving for y, we get $y = (-1/2) x + 11/10$. $m = -1/2$. $m_\perp = 2$. So $y' =$ $2x + 4 = 2$; $x = -1$; $y = (-1)^2 + 4(-1) + 7 = 4$.

The equation is $2 = \dfrac{y - 4}{x + 1}$.

E. A tough problem. If you draw the graph, (5, 9) is *not* on the parabola. There are two lines.

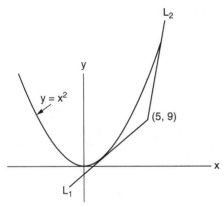

The slope, the derivative, equals 2x. The equation of the line with slope $m = \dfrac{y - 9}{x - 5}$ becomes $2x = \dfrac{y - 9}{x - 5}$. But on this curve $y = x^2$.

The equation becomes $2x = \dfrac{x^2 - 9}{x - 5}$. Simplifying, we get $x^2 - 10x + 9 = (x - 9)(x - 1) = 0$.
$x = 1$ and 9. For $x = 1$, $m = 2x = 2$; for $x = 9$, $m = 2x = 18$.

The equations are $2 = \dfrac{y - 9}{x - 5}$ and $18 = \dfrac{y - 9}{x - 5}$.

F. A really tough problem. Read the answer to E before you try F. See figure on the next page.

A point on the graph $y = x^2$ is given by any point (a, a^2).

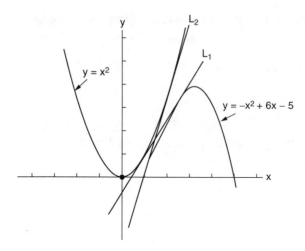

To be on the graph $y = -x^2 + 6x - 5$, it must be of the form $(b, -b^2 + 6b - 5)$.

The slopes have to be equal; $m = 2x = 2a$ has to be the same as $m = -2x + 6 = -2b + 6$.

$$2a = -2b + 6 \qquad \text{or} \qquad b = 3 - a$$

Also the slope has to be the same as the line joining these two points. The slopes involving derivatives have to be the same as the slopes involving the actual points. This is the toughest part of the problem to actually see yourself.

$$m = \frac{y_2 - y_1}{x_2 - x_1} \qquad 2a = \frac{(-b^2 + 6b - 5) - a^2}{b - a} \qquad b = 3 - a$$

Substituting, we get

$$2a = \frac{-(3 - a)^2 + 6(3 - a) - 5 - a^2}{3 - a - a}$$

Doing quite a bit of algebra, we get $2a^2 - 6a + 4 = 2(a - 2)(a - 1) = 0$.

If $a = 1$, $(a, a^2) = (1, 1)$. Since $b = a - 3$, $b = 2$.
$(b, -b^2 + 6b - 5) = (2, 3)$.

If a = 1 the slope $= 2a = 4b - 6 = \dfrac{3 - 1}{2 - 1} = 2$.

The equation of this line is $2 = \dfrac{y - 1}{x - 1}$ or $2 = \dfrac{y - 3}{x - 2}$.

If $a = 2$, $(a, a^2) = (2, 4)$; $b = 1$ and $(b, -b^2 + 6b - 5) = (1, 0)$.

The slope $= 2a = -2b + 6 = \dfrac{0 - 4}{1 - 2} = 4$. The equation of line 2 is $4 = \dfrac{y - 4}{x - 2}$ or $4 = \dfrac{y - 0}{x - 1}$.

Wow!!!!!

Let's try one more.

EXAMPLE 22—

Let

$$f(x) = \begin{cases} 3x + 4 & x > 0 \\ ax + b & x \le 0 \end{cases}$$

If f(x) is differentiable at x = 0, what does a + b equal?

If f(x) is differentiable, it first means that f(x) is continuous at x = 0.

$\lim\limits_{x \to 0^-}(3x + 4) = \lim\limits_{x \to 0^+}(ax + b)$. So b = 4. Also $f'(0^-) = f'(0^+)$. So a = 3. Therefore a + b = 7.

NOTE

This is the way the AP asked questions, only most are multiple choices.

I know I said one more, but let's try one more again.

EXAMPLE 23—

The line y = mx + b, b ≥ 4, is tangent to the curve f(x) = $-4(x - 4)^2 + 4$ at a point in the first quadrant. Find all possible values for b.

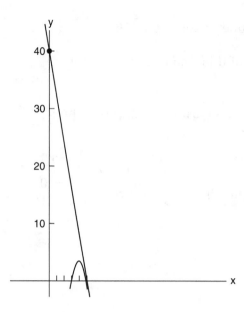

The tangent line has the highest y intercept when the graph of f(x) almost hits the x axis. You can see that f(5) = 0.

$f'(x) = -8(x - 4)$. $f'(5) = -8$. So m = -8; $y = mx + b$, or $0 = -8(5) + b$. So b is almost 40. The answer is $4 \leq b < 40$. [Remember (5, 0) is on the axis, not in the first quadrant!]

ANTIDERIVATIVES AND DEFINITE INTEGRALS

I put this section here for two reasons.

First, from a learning point, this section is necessary and pretty easy, and it should be here. Second, I am honoring the author of perhaps the best calculus book ever written, Abraham Schwartz, and his book *Calculus and Analytic Geometry*, editions 2 and 3. Professor Schwartz had the ridiculous idea that calculus should be presented in a way that students could actually learn and understand the material.

I have followed his example with a more relaxed teaching style.

Let's begin the section.

We are interested in the *antiderivative.* That is, given a function f(x), the antiderivative of f(x), called F(x), is a function such that F'(x) = f(x). We are going to explore methods of getting F(x).

The big problem in antiderivatives is that there is no product rule and no quotient rule. You might say, "Hooray! No rules to remember!" In fact, this makes antiderivatives much more difficult to find, and for many functions we are unable to take the antiderivatives. However, in Calc I, antiderivatives are very gentle. Only later do they get longer and more difficult.

RULE I

If F'(x) = G'(x), then F(x) = G(x) + C. If the derivatives are equal, the original functions differ by a constant. In other words, if you have one antiderivative, you have them all; you just have to add a constant.

EXAMPLE 24 SORT OF—

$$F(x) = x^2 + 7 \qquad G(x) = x^2 - 3$$

F'(x) = 2x = G'(x). The difference between F(x) and G(x) is a constant, 10.

RULE 2

If $f'(x) = x^N$ with $N \neq -1$, then

$$f(x) = \frac{x^{N+1}}{N + 1} + C$$

RULE 3

If $y' = kz'$, then $y = kz + C$.

Note I: If you notice, I didn't say to get the first edition. It was not very good. I have always said that he must have had divine inspiration for the next edition. I mention this to all my students. A female student once went home to find a friend of hers ready to throw out the book. She asked for and received the book. This student told me that Professor Schwartz's book read like a novel.

Well, no calculus book reads like a novel, but you understand how much she thought of the book.

Add I to the original exponent and divide by the new exponent plus a constant.

RULE 4

If $y' = f' + g'$, then $y = f + g + C$.

EXAMPLE 25—

$$\frac{dy}{dx} = 3x^4 - 8x^3 + 3x + 5 \qquad (5 = 5x^0)$$

$$y = \frac{3x^5}{5} - \frac{8x^4}{4} + \frac{3x^2}{2} + \frac{5x^1}{1} + C$$

$$y = \frac{3x^5}{5} - 2x^4 + \frac{3x^2}{2} + 5x + C$$

EXAMPLE 26—

$$\frac{dy}{dt} = \frac{4}{\sqrt{t}} - \frac{8}{t^5} + a = 4t^{-1/2} - 8t^{-5} + a \qquad (a = \text{constant})$$

$$y = \frac{4t^{1/2}}{^1/_2} - \frac{8t^{-4}}{-4} + at + C$$

$$= 8t^{1/2} + \frac{2}{t^4} + at + C$$

EXAMPLE 27—

If $y' = 3x^2 + 4x$ and $y = 2$ when $x = 3$, find y.

$$y = \frac{3x^3}{3} + \frac{4x^2}{2} + C = x^3 + 2x^2 + C$$

Now use $x = 3$, $y = 2$ to find C.

$$2 = (3)^3 + 2(3)^2 + C$$

$$2 = 27 + 18 + C$$

So $C = -43$.

$$y = x^3 + 2x^2 - 43$$

RULE 5

If $y' = u^n \dfrac{du}{dx}$ with $n \neq -1$, then $y = \dfrac{u^{n+1}}{n+1} + C.$

NOTE

Put $u = x$ in rule 5 and you will have rule 2. We will discover that when $N = -1$, the antiderivative is a logarithm.

EXAMPLE 28—

Find y if $y' = x(x^2 + 1)^{99}$.

We could multiply this out. On the grounds of sanity, we reject this technique.

Let $u = (x^2 + 1)$. $du/dx = 2x$.

$$y' = x(x^2 + 1)^{99} = \frac{1}{2}(x^2 + 1)^{99}(2x) = \frac{1}{2}u^{99}\frac{du}{dx}$$

$$y = \frac{1}{2}\left(\frac{u^{100}}{100}\right) + C = \frac{(x^2 + 1)^{100}}{200} + C$$

EXAMPLE 29—

If $\dfrac{dy}{dx} = \dfrac{x^3}{(x^4 + 11)^6}$, find y.

Let $u = x^4 + 11$. $du/dx = 4x^3$.

$$\frac{dy}{dx} = \frac{1}{4}(x^4 + 11)^{-6}(4x)^3 = \frac{1}{4}u^{-6}\frac{du}{dx}$$

$$y = \frac{1}{4}\left(\frac{u^{-5}}{-5}\right) + C = \frac{-1}{20(x^4 + 11)^5} + C$$

EXAMPLE 30—

If $v = 6t^2 + 4t + 3$ and $s = 40$ when $t = 1$, find s; $s = $ distance, $v = $ velocity, $t = $ time, $v = ds/dt$.

$$\frac{ds}{dt} = 6t^2 + 4t + 3$$

$$s = \frac{6t^3}{3} + \frac{4t^2}{2} + 3t + C$$

$$s = 2t^3 + 2t^2 + 3t + C$$

Use $t = 1$ when $s = 40$ to get $40 = 2(1)^3 + 2(1)^2 + 3(1) + C$. $C = 33$.

$$s = 2t^3 + 2t^2 + 3t + 33$$

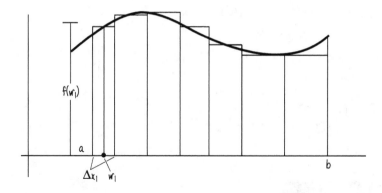

The development of the area as motivation for the definite integral is detailed in most calculus books. We will sketch the development.

A. Given the region $y = f(x)$, $x = a$, $x = b$, x axis.

B. Divide the interval $[a, b]$ into n intervals Δx_1, Δx_2, Δx_3, ..., Δx_n. Here Δx_i represents an arbitrary interval.

C. Let w_i be any point in the interval Δx_i.

D. Δx_1 represents the width of the first approximating rectangle. $f(w_1)$ represents the height of the first rectangle. $\Delta x_1 \, f(w_1)$ represents the area of the first approximating rectangle.

E. We do this for all the rectangles. We get

$$f(w_1) \, \Delta x_1 + f(w_2) \, \Delta x_2 + f(w_3) \, \Delta x_3 + \cdots$$

$$+ \, f(w_n) \, \Delta x_n = \sum_{i=1}^{n} f(w_i) \, \Delta x_i$$

$$\lim_{\substack{x \to \infty \\ \text{all } \Delta x_i \to \infty}} \sum_{i=1}^{n} f(w_i) \, \Delta x_i = \int_{a}^{b} f(x) \, dx$$

the definite integral.

In Example 29, the definite integral represents the area. As in the case of derivatives, we would like the rules so that it would be unnecessary to do this process of approximating rectangles.

NOTE

If the velocity $= v(t)$, then

$$\int_{a}^{b} v(t) dt$$

represents the distance traveled from time a to time b.

RULE 6

$$\int_{a}^{b} 1 dx = b - a$$

RULE 7

$$\int_{a}^{b} cf(x) = c \int_{a}^{b} f(x) \, dx$$

RULE 8

$$\int_a^b f(x) + g(x)\, dx = \int_a^b f(x)\, dx + \int_a^b g(x)\, dx$$

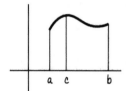

RULE 9

$$\int_a^b f(x)\, dx = \int_a^c f(x)\, dx + \int_c^b f(x)\, dx$$

RULE 10

$$\int_a^a f(x)\, dx = 0$$

RULE 11

$$\int_a^b f(x)\, dx = -\int_b^a f(x)\, dx$$

RULE 12

Fundamental theorem of calculus.

In the spirit of Professor Schwartz, we will demonstrate the fundamental theorem.

Let $f(x)$ be continuous on $a \le x \le b$. Let $F(x)$ exist such that $F'(x) = f(x)$.

Then $\int_a^b f(x)\, dx = F(b) - F(a)$.

Part 1: Let $A(x) = \int_a^x f(x)\, dx$.

Let's take the derivative of A(x).

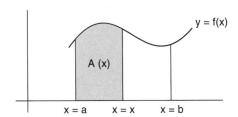

$\lim\limits_{\Delta x \to 0} \dfrac{A(x + \Delta x) - A(x)}{\Delta x}$. We have

$$\dfrac{\displaystyle\int_{a}^{x+\Delta x} f(x)\,dx - \int_{a}^{x} f(x)\,dx}{\Delta x} = \dfrac{\Delta A}{\Delta x}$$

Pictorially, it looks like this:

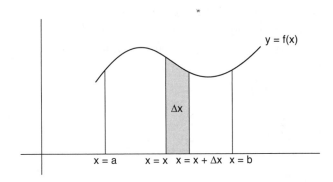

As Δx goes to 0, ΔA approximates a rectangle. In the limit,

$$\dfrac{\Delta A}{\Delta x} = \dfrac{\text{area}}{\text{base}} = \text{height} = f(x)$$

In other words, the derivative of A(x), the *area*, is f(x).

Part 2: Both $A(x)$ and $F(x)$ are antiderivatives of $f(x)$. By rule 1, $A(x) = F(x) + C$.

$A(a) = F(a) + C = 0$. (See the picture—the left and right ends are the same!)

$C = -F(a)$. $A(x) = F(x) - F(a)$.

The area of the whole region

$$A(b) = \int_a^b f(x)\, dx = F(b) - F(a).$$ (Again look at the picture.)

In words, the area under the curve is any antiderivative evaluated at the right end minus the same antiderivative evaluated at the left end.

NOTE

A rigorous proof of this theorem as well as rule 1 and other integral theorems in this book involves the mean value theorem that we discuss later.

We will not do a rigorous proof (not difficult) because this one gives you much more of an idea of what is going on.

Let's do examples.

EXAMPLE 31

$$\int_1^4 (3x^2 + 6)\, dx$$

$$= \int_1^4 3x^2\, dx + 6 \int_1^4 1\, dx$$

$$= x^3 + 6x \Big|_1^4 = [4^3 + 6(4)] - [1^3 + 6(1)] = 81$$

NOTE

$\int_a^b f(x)\,dx$, the definite integral, is a number if a and b are numbers. $\int f(x)\,d(x)$, the antiderivative or indefinite integral, is a family of functions, each of which differs from the others by a constant.

   We rewrite previous rules, using the indefinite integral.

$$\int x^n\,dx = \frac{x^{n+1}}{n+1} + C \qquad n \neq -1$$

$$\int [cf(x) + g(x)]\,dx = c\int f(x)\,dx + \int g(x)\,dx$$

$$\int u^N \frac{du}{dx} \cdot dx = \frac{u^{N+1}}{N+1} + C \qquad N \neq -1$$

EXAMPLE 32—

$$\int \frac{(x^4 + 3x)^2}{x^5}\,dx$$

$$\frac{(x^4 + 3x)^2}{x^5} = \frac{x^8 + 6x^5 + 9x^2}{x^5} = \frac{x^8}{x^5} + \frac{6x^5}{x^5} + \frac{9x^2}{x^5}$$

$$= x^3 + 6 + 9x^{-3}$$

$$\int \frac{(x^4 + 3x)^2\,dx}{x^5} = \int (x^3 + 6 + 9x^{-3})\,dx = \frac{x^4}{4} + 6x + \frac{9x^{-2}}{-2} + C$$

$$= \frac{x^4}{4} + 6x - \frac{9}{2x^2} + C$$

EXAMPLE 33—

$$\int_0^3 x^2(x^3 + 1)^{1/2}\,dx$$

This is a change of variables. We let u equal whatever is in parentheses, under a radical sign, etc. Let u = x³ + 1. du/dx = 3x². Solve for dx. du = 3x² dx. So dx = du/3x².

The limits must change. $u = x^3 + 1$. $x = 0$. $u = 0^3 + 1 = 1$.
$$x = 3. \ u = 3^3 + 1 = 28.$$

$$\int_{x=0}^{x=3} x^2(x^3 + 1)^{1/2}\,dx = \int x^2 u^{1/2}\,\frac{du}{3x^2}$$

$$= \int_{u=1}^{u=28} \left(\frac{1}{3}\right) u^{1/2}\,du$$

$$= \frac{1}{3}\,u^{3/2}\,\frac{3}{2}\,\Big|_1^{28}$$

$$= \left(\frac{2}{9}\right) u^{3/2}\,\Big|_1^{28} = \left(\frac{2}{9}\right)(28^{3/2} - 1)$$

EXAMPLE 34—

Here's a tricky one.

$$\int_1^2 \sqrt{x^4 - x^2}\,dx \qquad \sqrt{x^4 - x^2} = \sqrt{x^2(x^2 - 1)} = x\sqrt{x^2 - 1}$$

Let $u = x^2 - 1$. $x = 1; u = 0$. $x = 2; u = 3$. $dx = du/2x$.

$$\int_{x=1}^{x=2} \sqrt{x^4 - x^2}\,dx = \int_{x=1}^{x=2} x(x^2 - 1)^{1/2}dx = \int x u^{1/2}\,\frac{du}{2x}$$

$$= \int_{u=0}^{u=3} \frac{1}{2}\,u^{1/2}du = \frac{1}{2}\,\frac{u^{3/2}}{3/2}\,\Big|_{u=0}^{u=3}$$

$$= \left(\frac{1}{3}\right) u^{3/2}\,\Big|_{u=0}^{u=3} = \left(\frac{1}{3}\right)3^{3/2} - 0$$

$$= \left(\frac{1}{3}\right)(3)(3)^{1/2} = 3^{1/2}$$

NOTE

If you make a u substitution and the x's do not cancel, either a new technique must be used, or the problem cannot be solved.

EXAMPLE 35—

$$\int \sin 5x \, dx$$

Let $u = 5x$. $dx = \dfrac{du}{5}$.

$$\int \sin 5x \, dx = \int \sin u \, \frac{du}{5}$$

$$= \left(\frac{1}{5}\right) \int \sin u \, du = \left(\frac{1}{5}\right)(-\cos u) + C$$

$$= \frac{-\cos 5x}{5} + C$$

The integral of the sine is minus the cosine, and the integral of the cosine is the sine.

You must replace u with 5x since the problem originally had x.

NOTE

You must know how to do an integral like this one by sight, because if you don't, some of the Calc II integrals will come virtually endless.

EXAMPLE 36—

$$\int_{x=0}^{\pi/12} \tan^4 3x \sec^2 3x \, dx$$

You must remember that the derivative of the tangent is the secant squared.

Let $u = \tan 3x$. $du = 3 \sec^2 3x \, dx$. $du/3 = \sec^2 3x \, dx$.
$x = 0$; $\tan (0) = 0$. So $u = 0$.
$x = \pi/12$. $\tan (3\pi/12) = \tan (\pi/4) = 1$. So $u = 1$.

$$\int_{x=0}^{\pi/12} \tan^4 3x \sec^2 3x \, dx = \int_{u=0}^{u=1} \left(\frac{1}{3}\right) u^4 \, du$$

$$= \left(\frac{1}{3}\right)\frac{u^5}{5}\Big[_0^1 = \left(\frac{1}{15}\right) u^5 \Big[_0^1 = \frac{1}{15}$$

Finding the Area under the Curve by Using the Definition of the Definite Integral

One of the most laborious tasks is to find the area by using the definition. Doing one of these problems will make you forever grateful that there are some rules for antiderivatives, especially the fundamental theorem of calculus.

EXAMPLE 37—

Find $\int_3^6 (x^2 + 4x + 7) \, dx$ by using the definition of the definite integral.

Before we start, we need two facts:

$$1 + 2 + 3 + \ldots + n = n(n + 1)/2$$

$$1^2 + 2^2 + 3^2 + \ldots + n^2 = n(n + 1)(2n + 1)/6$$

These formulas for the sum of the first n positive integers and the sum of the squares of the first n positive integers can be found in some precalculus books, but are not too easily proved.

Now we are ready—not happy—but ready to start.

1. Divide the interval $3 \leqslant x \leqslant 6$ into n equal parts. At the left end $3 = x_0$, $x_1 = 3 + 1 \, \Delta x$, $x_2 = 3 + 2 \, \Delta x$, $x_3 = 3 + 3 \, \Delta x, \ldots$, and $x_n = 3 + n \, \Delta x = 6$. Solving for Δx gives $\Delta x = (6 - 3)/n = 3/n$.

2. From before, the approximate sum is $f(w_1) \, \Delta x_1 + f(w_2) \, \Delta x_2 + f(w_3) \, \Delta x_3 + \ldots f(w_n) \, \Delta x_n$. All the Δx_1 are equal to Δx, and we take the right end of each interval as the point where we will take the height. Therefore $w_1 = x_1$, $w_2 = x_2, \ldots$, and $w_n = x_n$.

3. Rewriting step 2, we factor out the Δx and get $[f(x_1) + f(x_2) + f(x_3) + \cdots + f(x_n)] \, \Delta x$.

4. Now $f(x) = x^2 + 4x + 7$.

    $f(x_1) = (3 + 1\,\Delta x)^2 + 4(3 + 1\,\Delta x) + 7$

    $f(x_2) = (3 + 2\,\Delta x)^2 + 4(3 + 2\,\Delta x) + 7$

    $f(x_3) = (3 + 3\,\Delta x)^2 + 4(3 + 3\,\Delta x) + 7$

    $f(x_n) = (3 + n\,\Delta x)^2 + 4(3 + n\,\Delta x) + 7$

5. Our task is now to add all these and then multiply everything by Δx.

A. If we multiply out $f(x_1)$, the number we get from this term is $3^2 + 4(3) + 7 = 28$. We see that every term, if we were to multiply them all out, would have a 28. Since there are n terms, the sum would be 28n.

B. Let's look at the Δx terms. $f(x_1)$ gives us $6(1\,\Delta x) + 4(1\,\Delta x) = 10(1\,\Delta x)$. $f(x_2)$ gives us $6(2\,\Delta x) + 4(2\,\Delta x) = 10(2\,\Delta x)$. Similarly, $f(x_3) = 10(3\,\Delta x)$. And $f(x_n) = 10(n\,\Delta x)$. Adding and factoring, we get $10\,\Delta x(1 + 2 + 3 + \cdots + n)$.

C. Looking at the $(\Delta x)^2$ terms, we would get

    $$1^2(\Delta x)^2 + 2^2(\Delta x)^2 + 3^2(\Delta x)^2 + \cdots + n^2(\Delta x)^2$$

    Factoring, we would get

    $$(\Delta x)^2(1^2 + 2^2 + 3^2 + \cdots + n^2)$$

6. Now, adding everything, multiplying by Δx, hoping everything fits on one line, we get

    $$[28n + 10\,\Delta x\,(1 + 2 + 3 + \cdots + n) + (\Delta x)^2 (1^2 + 2^2 + 3^2 + \cdots + n^2)]\,\Delta x$$

7. Substituting the formulas in the beginning and remembering $\Delta x = 3/n$, we get

    $$\{28n + 10(3/n)[n(n + 1)/2] + (3/n)^2\,[n(n + 1)(2n + 1)/6]\}\,(3/n)$$

8. Using the distributive law, we get three terms:

   A. $(28n)(3/n) = 84$

   B. $10(3/n)[n(n + 1)/2](3/n) = 45(n + 1)/n$
      $$\lim_{n \to \infty} B = 45(1) = 45$$

   C. $(9/n^2)[n(n + 1)(2n + 1)/6](3/n) = [(27/6)$
      $(2n^2 + 3n + 1)/6(n^2)$
      $$\lim_{x \to \infty} C = (27/6)(2) = 9$$

9. Adding $A + B + C$, we get a value for the area of
   $84 + 45 + 9 = 138$.

NOTE

By letting n go to infinity, we are doing two things: chopping up the interval $3 \leq x \leq 6$ into more and more rectangles and, since $\Delta x = 3/n$, making each rectangle narrower and narrower.

Wow!!!!!!!! Are we not grateful to for the fundamental theorem?

$$\int_3^6 (x^2 + 4x + 7)\, dx = \frac{x^3}{3} + 2x^2 + 7x \Big|_3^6$$

$$= \left[\frac{6^3}{3} + 2(6)^2 + 7(6)\right] - \left[\frac{3^3}{3} + 2(3)^2 + 7(3)\right]$$

$$= (72 + 72 + 42) - (9 + 18 + 21)$$

$$= 186 - 48 = 138$$

Finally we will learn, sadly, that most integrals cannot be solved. As we go on in math, we will learn many approximation methods. Also, we will learn how accurate these approximations are. This is OK because we don't live in a perfect world—or is this a surprise to you? As we go on, the methods of approximation will become more involved. Let us take a look at some crude ones. We will approximate

$$\int_{-1}^{5} (x^2 + 2)\, dx$$

in three ways.

EXAMPLE 38—

Approximate this integral with three equal subdivisions, using the right end of each one. Here's the picture:

The approximate area is $f(w_1)\,\Delta x_1 + f(w_2)\,\Delta x_2 + f(w_3)$

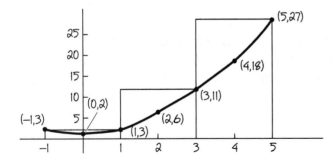

Δx_3. Each $x = 2$, and $w_1 = 1$, $w_2 = 3$, and $w_3 = 5$, the right ends of each interval. The approximation is $(\Delta x)\,[f(1) + f(3) + f(5)] = 2(3 + 11 + 27) = 82$.

EXAMPLE 39—

Same picture, same intervals, the minimum approximation, the smallest value in each interval S_3.

$S_3 = \Delta x\,[f(0) + f(1) + f(3)] = 2(2 + 3 + 11) = 32$

EXAMPLE 40—

Same picture, $x_0 = -1$, $x_1 = 0$, $x_2 = \frac{1}{2}$, $x_3 = 1$, $x_4 = 5$; midpoints.

The approximation is

$$f(w_1)\,\Delta x_1 + f(w_2)\,\Delta x_2 + f(w_3)\,\Delta x_3 + f(w_4)\,\Delta x_4$$

$$= f(-\tfrac{1}{2})\,(1) + f(\tfrac{1}{4})\,(\tfrac{1}{2}) + f(\tfrac{3}{4})\,(\tfrac{1}{2}) + f(3)(4)$$

$$= \left(\frac{9}{4}\right)(1) + \left(\frac{21}{16}\right)\left(\frac{1}{2}\right) + \left(\frac{29}{16}\right)\left(\frac{1}{2}\right) + 11(4)$$

$$= 48\,\frac{9}{16}$$

Of course we do not have to approximate this integral, since it is easily done. However, we would need to approximate, say, this one:

$$\int_{-1}^{5} \frac{1}{1 + x^4}\,dx$$

A theorem that is now mentioned much more often in Calc I is the *average value theorem*. It says that if we have an integrable function on the interval $a \leq x \leq b$, there exists a point c between a and b such that

$$f(c) = \frac{1}{b - a}\int_{a}^{b} f(x)\,dx$$

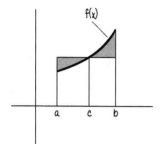

We will demonstrate by picture.

Suppose we have the function as pictured. There is a point c where the two shaded areas are the same. Fill in the top one in the bottom space. Thus the area of the rectangle equals the area under the curve. Area of the rectangle is base times height. Base = b − a. Height = f(c).

$$f(c)\,(b - a) = \int_{a}^{b} f(x)\,dx$$

Now divide by b − a.

That's it. Let's do an example.

EXAMPLE 41

Find the average value for $f(x) = x^2$, $2 \leq x \leq 5$.

$$\frac{1}{5-2} \int_{2}^{5} x^2 \, dx = \frac{x^3}{3}\bigg|_{2}^{5} = \frac{1}{9}(5^3 - 2^3) = \frac{1}{9}(117) = 13$$

If I actually wanted to find the point c in the theorem that gives the average value, $f(c) = c^2 = 13^{1/2} = 3.6$, which clearly is between 2 and 5.

Let's do a word problem.

EXAMPLE 42

During the 12 hours of daylight, the temperature in degrees Fahrenheit is give by $T = 60 + 4t - t^2/3$. Find the average temperature over the 12-hour period. The average temperature value is

$$= \frac{1}{12-0} \int_{0}^{12} \left(60 + 4t - \frac{t^2}{3}\right) dt$$

$$= \frac{1}{12}\left(60t + 2t^2 - \frac{t^3}{9}\right)$$

$$= \frac{60(12)}{12} + \frac{2(12)(12)}{12} - \frac{(12)(12)(12)}{(3)(3)(12)}$$

$$= 68°F \ (20°C), \text{ a delightful average temperature.}$$

ANSWER TO PROBLEM ON PAGE?!?!?

$2^\pi = 2^{3.141592653\cdots}$ is the limit of the following terms: 2^3, $2^{3.1}$, $2^{3.14}$, $2^{3.141}$, and so on.

As you may already have seen, answers in calculus aren't always nice, nice looking, or simple.

CHAPTER 3

CURVE SKETCHING MADE EASY

The topic I think I can teach better than anyone else in the world is this one. The only question was, "Could I write it down?" I think I did, and I think you'll really enjoy it!

Since we can spend an almost infinite amount of time on the topic, we will restrict our discussion to polynomials and rational functions (polynomials over polynomials), except for a few examples at the end.

TERMS AND SPECIAL NOTATIONS

1. For curve-sketching purposes, we define an *asymptote* as a line to which the curve gets very close at the end but never hits. All your life you have been told a curve cannot hit an asymptote. This is wrong. An asymptote is a straight-line approximation to a curve near its end, that is, when x or y goes to plus or minus infinity. In the middle of the curve, the curve is not a straight line and can hit the asymptote. The x axis is an asymptote although the curve hits the axis five

times. At the end of the curve, the curve gets close to the axis but does not hit it.

2. $|f(3)| = $ infinity. As x gets close to 3, f(x) gets very, very big (heading to plus infinity) or very, very small (heading to minus infinity).

3. $f(6^+)$. Substitute a number a little larger than 6, such as 6.01.

4. $f(6^-)$. Substitute a number a little smaller than 6, such as 5.99.

Our first goal is to sketch, in under 2 minutes, curves such as

$$y = f(x) = \frac{(x - 2)^6 (x + 1)^7}{(x + 2)^3 (x - 1)^{20}}$$

Yes, not only is it possible, but almost all my students do it and so will you !!!!!

INTERCEPTS

x intercepts. Just as on a straight line, an x intercept is a point where y = 0. If we have a fraction, y = 0 means that the top of the fraction = 0.

EXAMPLE 1

$$y = \frac{x^4(2x - 3)^7 (x + 4)^8}{(x + 2)^3 (x - 1)^{20}}$$

y = 0 means the top is 0. "Top is 0" means x = 0, 2x − 3 = 0, or x + 4 = 0. You ignore the exponents, since $x^4 = 0$ means x = 0. x = 0, 3/2, and −4. The intercepts are (0, 0), (1.5, 0), and (−4, 0).

EXAMPLE 2—

$$y = \frac{x^2 - 2x - 3}{x}$$

Factor the top.

$(x - 3)(x + 1) = 0$. Intercepts are $(3, 0)$ and $(-1, 0)$.

EXAMPLE 3—

$$y = x^4 (x - 1)^5 (x - 2)^6$$

Intercepts are $(0, 0)$, $(1, 0)$, and $(2, 0)$.

   y intercepts. Just as on the line, a y intercept is a point where $x = 0$.

EXAMPLE 4—

$$y = \frac{(x + 1)^8 (x - 2)^5}{(x - 1)^7 (x + 4)}$$

Substitute x = 0.

$\dfrac{(1)^8 (-2)^5}{(-1)^7 (4)} = 8$. y intercept is $(0, 8)$.

EXAMPLE 5—

$$y = \frac{x^2 - 2x - 3}{x^2}$$

For $x = 0$, we get $-3/0$. There is no y intercept.
   Warnings:

1. If you get the *sign* of the y intercept wrong, you will *never, never* sketch the curve properly.

2. Functions have one y intercept at most (one or none).

3. If we have the intercept $(0, 0)$, it is one of the x intercepts, maybe the only one, but the only y intercept. We do not have to waste time trying to find another one!

VERTICAL ASYMPTOTES

A rational function has a vertical asymptote whenever the bottom of the fraction is equal to 0.

EXAMPLE 6

$$y = \frac{x + 31}{x^4(x - 4)^6 (x + 3)^5}$$

Asymptotes are vertical lines $x = 0$, $x = 4$, and $x = -3$.

Two more definitions need to be reviewed before we move on.

DEFINITION A

Degree—if a polynomial has one variable, the degree is the highest exponent.

DEFINITION B

Leading coefficient—the coefficient of the highest power.

EXAMPLE 7

$$y = 4x^3 - 7x^6 + 2$$

Degree is 6. Leading coefficient is -7.

NOTE

Since polynomials have no denominators, they have no vertical asymptotes. As we will see, they have no asymptotes at all.

HORIZONTAL ASYMPTOTE TYPE I

(Don't be scared. There are only two types!)

Suppose $y = P(x)/Q(x)$. P and Q are polynomials. If the degree of P (top) is less than the degree of Q (bottom), the horizontal asymptote is $y = 0$, the x axis.

EXAMPLE 8

$$y = \frac{3x^2 - 7x}{8 + 5x^4} = \frac{3x^2/x^4 - 7x/x^4}{8/x^4 + 5x^4/x^4} = \frac{3/x^2 - 7/x^3}{8/x^4 + 5}$$

As x goes to infinity, $3/x^2$, $-7/x^3$, and $8/x^4$ all go to 0. So $y = (0 - 0)/(0 + 5) = 0$. The asymptote is $y = 0$!!!

NOTE I
We divided by x^4, that is, x to the highest power.

NOTE 2
When you do the problem, don't actually do this. Since you know that anytime the degree of the top is smaller than the degree of the bottom you get $y = 0$, just use $y = 0$ when this happens. Easy, isn't it?!!

HORIZONTAL ASYMPTOTE TYPE 2

EXAMPLE 9—

$$y = \frac{6x^3 + 2}{5x - 7x^3}$$

Both degrees are 3. If the degree of the top equals the degree of the bottom, the horizontal asymptote is $y = a/b$, where a is the leading coefficient of the top and b is the leading coefficient of the bottom. Asymptote is $y = 6/(-7)$. Let us verify.

$$y = \frac{6x^3 + 2}{5x - 7x^3} = \frac{(6x^3)/(x^3) + 2/(x^3)}{(5x)/(x^3) - (7x^3)/(x^3)} = \frac{6 + 2/(x^3)}{(5)/(x^2) - 7}$$

As x goes to infinity, $2/x^3$ and $5/x^2$ go to 0. Asymptote is $y = -6/7$.

OBLIQUE (SLANTED LINE) ASYMPTOTE

This occurs when the degree of the top is exactly 1 more than the degree of the bottom.

EXAMPLE 10

$$y = \frac{x^2 - 2x - 3}{x + 4}$$

Degree of the top = 2; degree of the bottom = 1. It is an oblique asymptote.

We must, unfortunately, long-divide the bottom into the top. If you know it, use synthetic division.

$$x + 4 \overline{)x^2 - 2x - 3} \quad \frac{x - 6 + 21/(x+4)}{} \qquad y = \frac{x^2 - 2x - 3}{x+4} = x - 6 + \frac{21}{x+4}$$
$$\underline{x^2 + 4x}$$
$$-6x - 3$$
$$\underline{-6x - 24}$$
$$21$$

As x goes to infinity, the remainder $21/(x + 4)$ goes to 0. The oblique asymptote is $y = x - 6$.

NOTE I

If the degree of the top is more than the degree of the bottom but is not 1, there are no oblique asymptotes.

NOTE 2

At most there is one oblique asymptote or one horizontal asymptote, but not both. There might be neither.

CURVE SKETCHING BY THE PIECES

Before we undertake along examples, we will examine each piece. When you understand each piece, the whole will be easy.

EXAMPLE 11

$$f(x) = 3(x - 4)^6$$

The intercept is (4, 0). We would like to know what the curve looks like near (4, 0). Except at the point (4, 0), we do not care what the exact value is for y, which is necessary in an exact graph. In a sketch we are interested in only the *sign* of the y values. We know f(4) = 0. f(3.9) = $3(3.9 - 4)^6$ = 0.000003. We don't care about its value. We only care that it is positive. Our notation will be f(4⁻) is positive. Similarly f(4⁺) is positive. What must the picture look like? At x = 4, the value is 0. To the left and to the right, f(x) is positive.

The picture looks like this where a = f(4⁻) and b = f(4⁺):

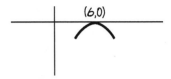

EXAMPLE 12—

$f(x) = -7(x - 6)^{100}$

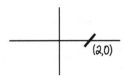

Intercept is (6, 0). f(6⁻) is negative. f(6⁺) is negative. The picture is . . .

To summarize, if the exponent is *even* and *positive,* the sketch does *not* cross at the intercept.

EXAMPLE 13—

$f(x) = 6(x - 2)^{11}$

Intercept is (2, 0). f(2⁻) is *negative*; f(2⁺) is *positive*. The curve around (2, 0) looks like . . .

EXAMPLE 14—

$f(x) = -7(x + 4)^{3743}$

Intercept is (−4, 0), f(−4⁻) is *positive*; f(−4⁺) is *negative.* Around (−4, 0) the curve will look like . . .

To summarize, if the exponent is *odd* and *positive,* the sketch *will* cross at the intercept.

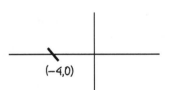

Let's see what it will look like near the vertical asymptotes. We attack the problem in exactly the same way.

EXAMPLE 15—

$f(x) = \dfrac{7}{(x - 4)^8}$

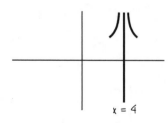

Vertical asymptote is at x = 4, f(4⁻) and f(4⁺) are positive, and the curve near x = 4 looks like . . .

To reemphasize, f(4⁺) = f(4.1) = 7/(4.1 − 4)⁸ = 7/0.00000001 = 700,000,000, which is *big*. The curve tends to plus infinity, from the right side of 4. Similarly, the curve goes to plus infinity from the left side.

EXAMPLE 16—

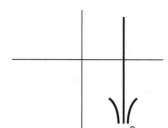

$$f(x) = \frac{-7}{(x - 3)^{100}}$$

Vertical asymptote is at x = 3. f(3⁻) and f(3⁺) are negative. The curve near x = 3 looks like . . .

To summarize, if the exponent is *even* and *positive* in the denominator, near the vertical asymptote, *both ends go to either plus infinity or minus infinity.*

EXAMPLE 17—

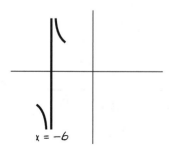

$$f(x) = \frac{9}{(x + 6)^{11}}$$

Asymptote is x = −6. f(−6⁻) is negative; f(−6⁺) is positive. The curve is . . .

EXAMPLE 18—

$$f(x) = -3/(x - 5)$$

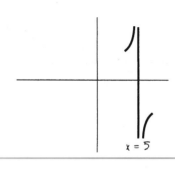

Exponent is 1 (odd). Asymptote is x = 5. f(5⁻) is positive; f(5⁺) is negative. The curve near x = 5 is . . .

To summarize, if the exponent is *odd* and *positive* in the denominator, on one side of the asymptote the curve goes to plus infinity and on the other the curve goes to minus infinity.

We are now ready to put all the pieces together. With some study and a little practice, you positively will be able to sketch curves with intercepts and asymptotes only in under two minutes!!!!!!!

EXAMPLE 19—

$$f(x) = \frac{(x - 2)^6 (x + 1)^7}{(x + 2)^3 (x - 1)^{20}}$$

First locate the intercepts. x intercept means y = 0
which means top of fraction = 0. (2, 0) and (−1, 0).
y intercept, the only long part. Substitute x = 0.

$$f(0) = \frac{(-2)^6(1)^7}{(2)^3(-1)^{20}} = 8$$

The intercept is (0, 8). Vertical asymptotes: bottom of
fraction = 0. x = −2 and x = 1. Horizontal asymptote:
If we multiply out the top (something you would *never*
do), the highest power of x on top is x^{13}. On the bottom
is x^{23}. The degree of the top is less than the degree of
the bottom.

The horizontal asymptote is y = 0. Oblique asymptote:
There is none, since there is a horizontal one.

We are ready to start the sketch. It is advisable to use
three colors—one for the axes, one for the asymptotes,
and one for the sketch.

We need to substitute only one number!!! That number is to the right of the rightmost vertical asymptote or
x intercept. $f(2^+)$ is positive. Since the power of (x − 2)
is even, namely, 6, the curve does not cross at (2, 0).
So far the sketch is . . .

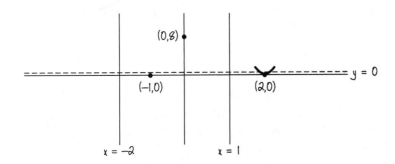

The sketch now heads for the asymptote x = 1. It must go to plus infinity, since if it went to minus infinity, there would have to be another x intercept between 1 and 2, and there isn't. The power of $(x - 1)^{20}$ is even. For an asymptote, that means both ends are in the same location. Since one part is at plus infinity, so is the other. The curve now heads through (0, 8) toward (−1, 0). It looks like . . .

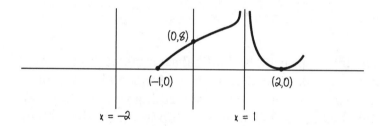

Since the power of (x + 1) is odd, 7, there is a crossing at (−1, 0). The sketch heads to minus infinity at x = −2. Since the power of (x + 2) is odd and an odd power means one end at plus infinity and the other end at minus infinity, one end is already at minus infinity, so that on the other side the curve goes to plus infinity. *Remember* both ends head for the horizontal asymptote y = 0, the x axis. The sketch can now be finished. . . .

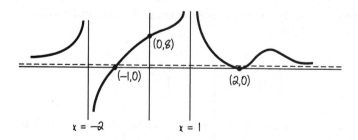

Well, that's it. With a little practice you'll be like a pro! We'll do more examples, of course. You should practice the ones in this book.

Before we do other examples, let us look at the right end of the curve that we drew above.

In the top figure, can we find out how high M is? Yes! Later in this section we will.

In the middle, could this be the right end? Perhaps, but we don't have enough info to know what the end looks like yet. In some very complicated cases, we might not ever be able to determine what the end is like (exactly).

In the bottom, could this be the right end? No!! There would have to be another intercept after (2, 0), and we know there aren't any.

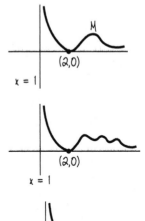

EXAMPLE 20—

$$f(x) = \frac{3x^4(x - 4)^5 (x - 8)^6}{2(x + 3)^7 (x + 6)^6 (x - 2)^2}$$

The x intercepts are (8, 0), (4, 0), and (0, 0) which is also the y intercept. Vertical asymptotes: $x = 2$, $x = -3$, $x = -6$. Horizontal asymptote: By inspection, the leading term on the top is $3x^{15}$, and the leading term on the bottom is $2x^{15}$. Same degree. The asymptote is $y = 3/2$, the leading coefficient on top over leading coefficient on bottom.

Starting the sketch at (8, 0), $f(8^+)$ is positive. The power of $(x - 8)$ is even, so there is no crossing. The sketch starts . . .

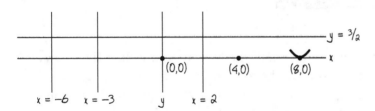

From the sketch, $f(4^+)$ is positive. Since $(x - 4)$ has odd power, there is a crossing at (4, 0) heading down to minus infinity at $x = 2$. The power of $(x - 2)^2$ is even. So one end at minus infinity means both ends are there, heading up to $(0, 0)$. x^4 is an even power, so that there is no crossing, and the sketch continues . . .

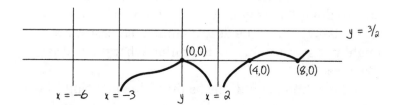

We head to minus infinity at $x = -3$. $(x + 3)^7$ is an odd power. So since one end is at minus infinity, the other is at plus infinity. It goes down (we don't know how far) but never hits the x axis and heads back to plus infinity at $x = -6$. Since $(x + 6)^6$ is an even power, both ends are at plus infinity. The *ends* both go to $y = \frac{3}{2}$, the horizontal asymptote. The sketch can now be finished . . . .

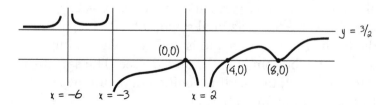

You should be getting better now. Let's try an oblique asymptote.

EXAMPLE 21—

$$f(x) = \frac{x^2 - 4x + 4}{x - 1}$$

Since the degree of the top is 1 more than the degree of the bottom, we have an oblique asymptote. We need three forms of the equation: the original, the factored, and the divided.

$$f(x) = \frac{x^2 - 4x + 4}{x - 1} = \frac{(x - 2)^2}{x - 1} = x - 3 + \frac{1}{x - 1}$$

x intercepts—top = 0 in the second form: (2, 0).
y intercept—x = 0, easiest found in the first form:
(0, −4). Vertical asymptote—first or second form: x = 1.
Oblique asymptote—third form: y = x − 3 with
remainder going to 0.

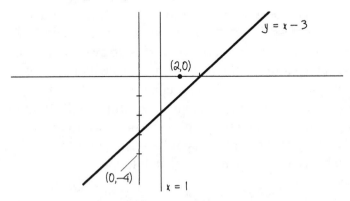

Again we look at the rightmost vertical asymptote or
x intercept, in this case (2, 0). f(2⁺) (form 2) is positive.
(x − 2)² is an even power, so there is no crossing, heading up to plus infinity at x = 1. Since the power of x − 1
(1) is odd, the other end is at minus infinity. The
sketch then goes through the point (0, −4) with both
ends going to the line y = x − 3. The sketch is . . .

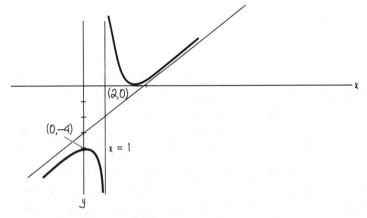

You should be getting *a lot better* now!! Let us try a
polynomial, to show you how easy it is.

EXAMPLE 22—

$f(x) = x^4(x - 1)^5 (x - 2)^6 (x - 3)^7 (5 - x)^9$

Intercepts: $(0, 0)$, $(1, 0)$, $(2, 0)$, $(3, 0)$, and $(5, 0)$. No asymptotes. Watch how easy this is! $f(5^+)$ is negative. $(5 - x)^9$ is an odd power, so cross. $(x - 3)^7$ is an odd power, so cross. $(x - 2)^6$ is an even power, no cross. $(x - 1)^5$ is an odd power, so cross. x^4 is an even power, no cross. The leading term, $-x^{31}$, dominates when x is big (say x = 100) or when x is small (say x = −100). $-(100)^{31}$ is negative—right end goes to minus infinity. $-(-100)^{31}$ is positive—left end goes to plus infinity. Briefly, right to left, looking at the exponents only—cross, cross, no cross, cross, no cross, and the sketch looks like . . .

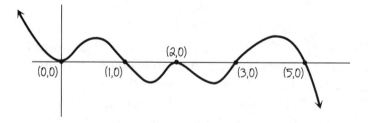

EXAMPLE 23—

$f(x) = \dfrac{(x - 4)^8}{x^3} = x^5 + \cdots$

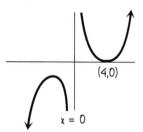

Intercept: $(4, 0)$. Vertical asymptote $x = 0$ (y axis). There is no oblique or horizontal asymptote since degree of top is 5 more than degree of bottom. $(100)^5$ is positive—right end goes to plus infinity. $(-100)^5$ is negative—left end goes to minus infinity. The sketch is . . . ta-da! . . .

The next area of curve sketching involves maximum points, minimum points, inflection points, and cusps. Since much of this involves derivatives and factoring, more care and time is needed.

EXAMPLE 24—

$$f(x) = \frac{(x - 1)(4x^2 - 9)}{(x - 1)(5x^2 + 30x + 45)}$$

List all values of x at which f(x) is not defined, all intercepts, and all asymptotes.

Factoring f(x), we get

$$f(x) = \frac{(x - 1)(2x - 3)(2x + 3)}{(x - 1)(5)(x + 3)(x + 3)}$$

It is not defined when the denominator is 0: when x equals 1, −3 (second equation).

The vertical asymptote is x = −3 (second equation).

(x = 1 is an open dot, where the curve is not defined.)

y intercept means x = 0 (first equation). We get (0, −9/45) = (0, −1/5).

x intercept means y = 0; numerator equals 0 (second equation): (3/2, 0), (−3/2, 0).

Again x = 1 is neither an x intercept nor a vertical asymptote.

In the first equation, degree of the top = degree of the bottom.

The horizontal asymptote is y = 4/5.

DEFINITIONS

Relative maximum (minimum)—the largest (smallest) y value in a region.

Absolute maximum (*minimum*)—the largest (smallest) y value of all.

If the sketch goes to plus infinity (minus infinity), all maximums (minimums) are relative. A, C, E, H, and J are relative maximums; B, D, F, G, I, and K are relative minimums. There are no absolute maximums. F is the absolute minimum.

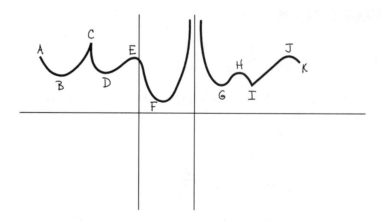

As you can see, there are three kinds of maximum and minimum.

1. If there is a finite domain, the left and right endpoints (A and K) are sometimes relative maximums or minimums. (They are in most of the examples we do.)

2. *Cusps,* C and I, will be discussed later.

3. *Round maximums* and *minimums* are discussed now. As you can see, the slope of the tangent line exists at these points and is equal to 0.

Let us proceed.

TESTING FOR ROUND MAXIMUMS AND MINIMUMS

If $y = f(x)$ and there is a round max or min at $x = c$, then $f'(c) = 0$.

Test 1 for round max and min points.

If $f'(c^-)$ is negative and $f'(c^+)$ is positive, there is a minimum.

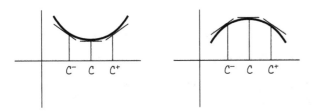

If $f'(c^-)$ is positive and $f'(c^+)$ is negative, there is a maximum.

If $f'(c^-)$ and $f'(c^+)$ are the same sign, it is neither a max nor a min.

Down   Up

Believe it or not, at this time we have to define the terms *down* and *up*. A curve is down (up) if the tangent line drawn to the curve is above (below) the curve itself.

DEFINITION

Inflection point—a point at which a curve goes from up to down or down to up.

Down to up   Up to down

At test for inflection points is $f''(c) = 0$. There are in fact two tests to tell whether a point is an inflection point.

TEST I (easier)

A. If $f'''(c) \neq 0$, then c is an inflection point.

B. If $f'''(c) = 0$, this test fails and you must use another test.

NOTE

If the third derivative is very difficult to find, use the other test.

TEST 2

A. If $f''(c^+)$ and $f''(c^-)$ have different signs, c is an inflection point.

B. If $f''(c^+)$ and $f''(c^-)$ have the same sign, c is not an inflection point.

There is also a second and easier test for max and min points. Suppose $f''(c) = 0$. If $f''(c)$ is positive, it means the slope is increasing. This means the curve is facing up, which means a *minimum*. Suppose $f'(c) = 0$ and $f''(c)$ is negative. This means the slope is decreasing, the curve faces down, and we have a *maximum*. If $f''(c) = 0$, then we use the other test.

PROBLEM

Before we sketch some more curves, let's make sure we all understand one another. There is a kind of problem my fellow lecturer Dan Mosenkis at CCNY likes to give his students. It's not my cup of tea or cup of anything else, but I think it will help you a lot. We have made up a craaaazy function, f(x). Its picture is on the next

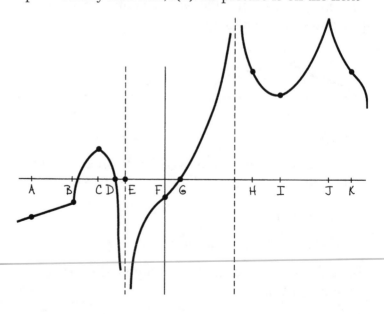

page. For each listed value of x, A through K, look at f(x) and determine the sign of f(x), f'(x), and f"(x) at each point. Enter one of the following symbols on the chart: + (if positive), − (if negative), 0, and ? (if it does not exist).

Answer at the end of this chapter!!

x	A	B	C	D	E	F	G	H	I	J	K
f(x)											
f'(x)											
f"(x)											

You are now ready for your first curve sketch involving max and min inflection points.

EXAMPLE 25—

$$y = f(x) = x^4 - 4x^3$$

Take three derivatives and factor each equation.

$$y = x^4 - 4x^3 = x^3(x - 4)$$

$$y' = 4x^3 - 12x^2 = 4x^2(x - 3)$$

$$y'' = 12x^2 - 24x = 12x(x - 2)$$

$$y''' = 24x - 24 = 24(x - 1)$$

Intercepts: y = 0 (0, 0), (4, 0). Max and min: $y' = 0. x = 0, x = 3$. To find the y values, we must substitute each x value *into the original equation,* since that is the curve we are sketching. We get the points (0, 0) and (3, −27).

NOTE

It is almost always easier to find the y value from the factored equation of y.

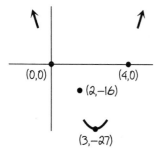

We will use both tests to test points. However, when you do the problem, once you get an answer for one point, go on to the next point.

We first substitute x = 0, 3 into f''(x) = 12x(x − 2). f''(0) = 0, so this test fails. f''(3) is positive, so (3, −27) is a minimum.

Test 2: f'(x) = 4x²(x − 3). f'(0⁺) is negative and f'(0⁻) is negative. (0, 0) is neither a max nor a min. f'(3⁻) is negative and f'(3⁺) is positive. Again, (3, −27) is a minimum.

Remember, always substitute into the factored form. When you are testing, do not evaluate. You are interested in only the sign of the answer.

Possible inflection points f''(x) = 0 = 12x(x − 2), x = 0, x = 2. Substituting back into the original equation (factored) for y, we get the points (0, 0) and (2, −16).

Let us use both tests.

Test 1: f'''(0) and f'''(2) are both not 0. Therefore (0, 0) and (2, −16) are inflection points.

Test 2: f''(0⁻) is positive and f''(0⁺) is negative. Different signs. (0, 0) is an inflection point. f''(2⁻) is negative and f''(2⁺) is positive. Different signs. (2, −16) is an inflection point.

What happens to the ends? For large x, y = (approximately) x⁴. f(100) is positive and f(−100) is positive. Both ends go to plus infinity.

Now connect the dots.

EXAMPLE 26—

$$f(x) = y = x^4 - 12x^2 = x^2(x^2 - 12)$$

$$= x^2(x - \sqrt{12})(x + \sqrt{12})$$

$$f'(x) = 4x^3 - 24x = 4x(x^2 - 6)$$

$$= 4x(x - \sqrt{6})(x + \sqrt{6})$$

$$f''(x) = 12x^2 - 24 = 12(x^2 - 2)$$

$$= 12(x - \sqrt{2})(x + \sqrt{2})$$

$$f''(x) = 24x$$

Intercepts: $y = 0$ $(0, 0)$, $(\sqrt{12}, 0)$, $(-\sqrt{12}, 0)$. Max, min possibles: $y' = 0$; $x = 0, \sqrt{6}, -\sqrt{6}$. Substituting for y values in the original equation, we get $(0, 0)$, $(\sqrt{6}, -36)$, $(-\sqrt{6}, -36)$. Testing, $f''(\sqrt{6})$ is positive. $(\sqrt{6}, -36)$ is a minimum. $f''(-\sqrt{6})$ is positive. $(\sqrt{6}, -36)$ is a minimum. $f''(0)$ is negative. $(0, 0)$ is a maximum.

Possible inflection points: $y'' = 0$. $x = \sqrt{2}, -\sqrt{2}$. Substituting for the y value into the original, we get the points $(\sqrt{2}, -20)$ and $(-\sqrt{2}, -20)$. Testing for inflection points, $f'''(\sqrt{2}), f'''(-\sqrt{2}) \neq 0$. Both $(\sqrt{2}, -20)$ and $(-\sqrt{2}, -20)$ are inflection points.

The ends $f(100)$ and $f(-100)$ are positive. Both ends go to plus infinity. The sketch in two stages is as follows:

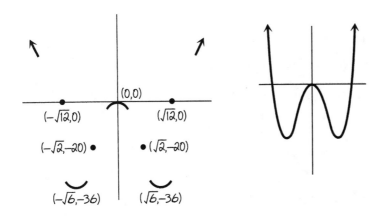

EXAMPLE 27—

$y = f(x) = 9 - x^2 \qquad -1 \leqslant x \leqslant 4$

$y = 9 - x^2 = (3 - x)(3 + x)$

$y' = -2x$

$y'' = -2$

$y''' = 0$

x intercept: $y = 0$ (3, 0)

NOTE

$x = -3$ is not in the domain.

y intercept: $x = 0$ (0, 9). Possible max, min: $y' = 0$, $x = 0$. We get the point (0, 9). (0, 9) is a maximum since y'' is always negative. There is no inflection point since y'' is never equal to 0.

Since the domain is finite, we must get values for the left and right ends of $x = -1$, $y = 8$ (−1, 8). $x = 4$, $y = -7$ (4, −7). We see that (4, −7) is an absolute minimum, (−1, 8) is a relative minimum, and (0, 9) is an absolute maximum.

We will examine several examples with fractional exponents. We might get cusps or a second kind of inflection point.

Given $y = f(x)$:

1. $|f'(c)| = $ infinity.

2. $f(c)$ exists [$f(c)$ is some number].

3. A. $f'(c^-)$ is negative and $f'(c^+)$ is positive, and the cusp looks like this:

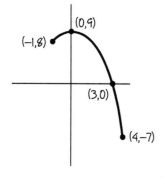

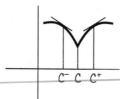

B. $f'(c^-)$ is positive and $f'(c^+)$ is negative, and the cusp looks like this:

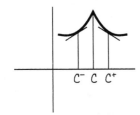

As we will see, if $f'(c^-)$ and $f'(c^+)$ have the same sign, we will get another kind of inflection point.

EXAMPLE 28—

$$y = f(x) = (x - 2)^{4/5} + 3$$

$$f'(x) = \left(\frac{4}{5}\right)(x - 2)^{-1/5} = \frac{4}{5(x - 2)^{1/5}}$$

$$f''(x) = \left(\frac{4}{5}\right)\left(-\frac{1}{5}\right)(x - 2)^{-6/5} = \frac{-4}{25(x - 2)^{6/5}}$$

We will test for the cusp first or second kind of inflection point.

1. $f'(2) = $ infinity.

2. $f(2) = 3\ (2, 3)$.

3. $f'(2^-)$ is negative. $f'(2^+)$ is positive. Cusp with the point down.

x intercept: $y = 0$. $-3 = (x - 2)^{4/5}$. $(x - 2)^{1/5} = \pm(-3)^{1/4}$, which is imaginary. No x intercepts. y intercept: $x = 0$. $y = (-2)^{4/5} + 3$. $[0, (-2)^{4/5} + 3] = A$. No round max or min and no inflection points. $f(1000)$ is positive and $f(-1000)$ is positive. Both ends go to plus infinity. The sketch is . . .

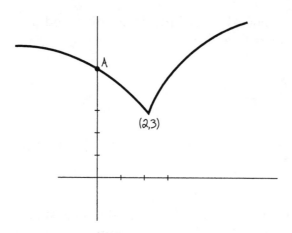

EXAMPLE 29—

$$y = f(x) = x^{1/3}$$

$$f'(x) = \frac{1}{3x^{2/3}} \qquad f''(x) = \frac{-2}{9x^{5/3}}$$

Intercept (0, 0). No round max or min and no inflection points.

1. $|f'(0)| = $ infinity.

2. $f(0) = 0$.

3. $f'(0^-)$, $f'(0^+)$ are both positive. Second kind of inflection point.

4. $f(100,000)$ is positive and $f(-100,000)$ is negative. The right end goes to plus infinity, and the left end goes to minus infinity.

The sketch:

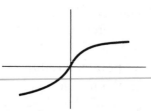

Instructors have been known to give messy examples. While I know yours would never give one, here are two that will take time.

EXAMPLE 30—

$$y = f(x) = \frac{4x}{x^2 + 1}$$

$$y' = 4\frac{(x^2 + 1)(1)-x(2x)}{(x^2 + 1)^2} = 4\frac{(1 - x^2)}{(x^2 + 1)^2}$$

$$y'' = 4\frac{(x^2 + 1)^2(-2x)-(1 - x^2)\,2(x^2 + 1)(2x)}{(x^2 + 1)^4}$$

$$= 4\frac{-2x(x^2 + 1)\big[(x^2 + 1) + 2(1 - x^2)\big]}{(x^2 + 1)^4} = \frac{-8x(3 - x^2)}{(x^2 + 1)^3}$$

No third derivative . . . too messy!

Intercept (0, 0). No vertical asymptote since $x^2 + 1 = 0$ has only imaginary roots. Horizontal asymptote is $y = 0$ since the degree of the top is less than the degree of the bottom.

Possible max, min: $y' = 0$. $1 - x^2 = 0$, $x = \pm 1$. Substituting in the original, we get (1, 2) and (−1, −2). $f''(1)$ is negative. (1, 2) is a maximum. $f''(-1)$ is positive. (−1, −2) is a minimum.

Possible inflection points: $y'' = 0$, $-8x(3 - x^2) = 0$. $x = 0, \pm\sqrt{3}$. Substituting into the original, we get the points (0, 0), $(\sqrt{3}, \sqrt{3})$, $(-\sqrt{3}, -\sqrt{3})$.

By the messier test, all three are inflection points.

If you liked that one, you'll love this one.

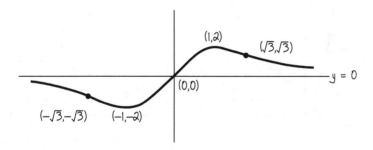

EXAMPLE 31

$-1 \leq x \leq 5$ $y = x^{2/3}(x^2 - 8x + 16) = x^{2/3}(x - 4)^2$

$$= x^{8/3} - 8x^{5/3} + 16x^{2/3}$$

$$y' = \left(\frac{8}{3}\right)x^{5/3} - \left(\frac{40}{3}\right)x^{2/3} + \left(\frac{32}{3}\right)x^{-1/3}$$

$$= \left(\frac{8}{3}\right)x^{-1/3}(x^2 - 5x + 4) = \frac{8(x - 4)(x - 1)}{3x^{1/3}}$$

$$y'' = \left(\frac{40}{9}\right)x^{2/3} - \left(\frac{80}{9}\right)x^{-1/3} - \left(\frac{32}{9}\right)x^{-4/3} = \frac{8(5x^2 - 10x - 4)}{9x^{4/3}}$$

There are no asymptotes. Intercepts: $y = 0$ (0, 0), (4, 0).
$y' = 0$, $x = 1, 4$. Substituting in the original, we get
(1, 9), (4, 0). $f''(1)$ is negative. (1, 9) is a maximum.
$f''(4)$ is positive. (4, 0) is a minimum.

Possible inflection points: $y'' = 0$. Using the quadratic formula and then substituting these values into
the original, we get approximately (2.34, 4.86) and
(−0.34, 9.23). These are both inflection points. Test for
cusps: $|f'(0)|$ is infinite. $f(0) = 0$. $f'(0^-)$ is negative and
$f'(0^+)$ is positive. Cusp points down. Left-end and
right-end substitution: $f(-1) = 25$ (−1, 25), $f(5) = 5^{2/3}$.
The point is approximately (5, 2.92).

Whew!!!! Let us sketch:

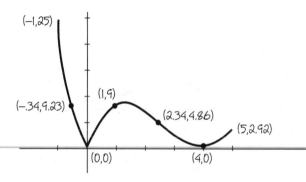

OTHER AIDS

Symmetry:

1. $f(x) = f(-x)$. Symmetric about the y axis.

2. $f(x) = -f(-x)$. Symmetric about the origin.

3. $g(y) = g(-y)$. Symmetric about the x axis. This is only found in graphing curves that are not functions of x.

EXAMPLE 32—

y axis symmetry:

$f(x) = x^2$

$f(-x) = (-x)^2 = x^2$

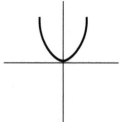

EXAMPLE 33—

Symmetric origin:

$f(x) = x^3$

$-f(-x) = -(-x)^3 = x^3$

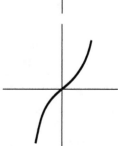

EXAMPLE 34—

x axis symmetry:

$g(y) = y^4$

$g(-y) = (-y)^4 = y^4$

For completeness, we will sketch a curve that is not a function of x or y.

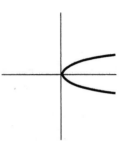

EXAMPLE 35—

$$y^2 = \frac{3x^2}{x^2 - 1} \qquad y = \pm\sqrt{\frac{3x^2}{x^2 - 1}}$$

Intercept is (0, 0). It turns out to be an isolated point since for all x values between -1 and 1, except $x = 0$,

$$\frac{x^2}{x^2 - 1}$$

is negative, making y imaginary.

Vertical asymptotes: $x = 1$, $x = -1$. Horizontal asymptotes: If we could talk about a degree, the degree of the bottom is 1 and the degree of the top is 1. Horizontal asymptote is $y = \pm\sqrt{3}$. Symmetry with respect to x, y axes. The sketch is as follows:

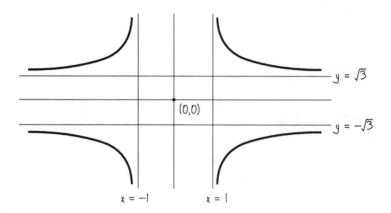

That's about it. You could write a whole book on curve sketching, a large book. I hope this gives you a solid idea on how to sketch curves.

Sometimes, you might like to know about the curve without actually drawing the curve itself. What questions may you ask? Well, I'll tell you.

EXAMPLE 36—

For $y = x^3 - 3x^2$, where is the function increasing and decreasing? What about its concavity?

y increases if y' is positive. Look at any graph that increases.

$$y' = 3x^2 - 6x = 3x(x - 2) > 0$$

Solving this quadratic inequality, we find that the curve increases if $x > 2$ or $x < 0$. That would mean y decreases when $0 < x < 2$.

Before, we found that a curve faces up if $y'' > 0$. $y'' = 6x - 6 > 0$ if $x > 1$, which is where the curve is up, and $y'' < 0$ if $x < 1$, where the curve is down.

You might try to do a sketch of this one on your own. If so, don't look at what follows.

Intercepts: $(0, 0)$, $(3, 0)$. Max, min: $(0, 0)$ is the max, $(2, -4)$ is the min. Inflection point: $(1, -2)$. Right end to plus infinity and left end to minus infinity. Polynomials have no asymptotes. And the sketch is . . .

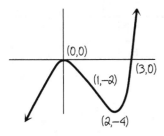

The AP exam might ask a question similar to this:

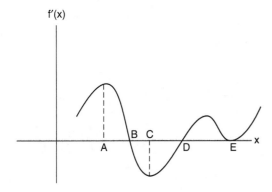

EXAMPLE 37—

At which point on the graph of f(x) is there a maximum? A B C D E

Notice that this is a graph of f'(x), not f(x)!!

The slope of f(x), or f'(x), is positive to the left of point B and negative to the right.

Point B is the answer, a maximum.

FYI point D is a minimum since the slope before D is negative and the slope after D is positive.

Point E is an inflection point since the slope before and after E is positive, while at E the slope is 0.

ANSWERS

Answers to the problem on page 91:

x	A	B	C	D	E	F	G	H	I	J	K
f(x)	−	−	+	0	?	−	0	+	+	+	+
f'(x)	+	?	0	−	?	+	+	−	0	?	−
f''(x)	0	?	−	−	?	0	+	+	+	?	0

CHAPTER 4

WORD PROBLEMS MADE EASY ··· WELL, LESS DIFFICULT

Word problems are not difficult because of the calculus. In fact, in more than 90 percent of word problems, the calculus is very easy. The problem is that few precalculus courses properly prepare you (or prepare you at all) for the setting up of the word problems. As you will see, the algebra is the difficult part of the problems.

The best we can do is to work out a number of word problems that are found in many calculus books. Many of the techniques involved in these problems are applicable to other word problems.

Because high schools tend to deemphasize word problems (to find a minimum or maximum or to find related rates), which I think, is a mistake, I have included only a portion of the problems for *Calc I for the Clueless*. They are the ones you most likely will see.

I must add that the person who reviewed and edited this book disagrees with me. She did a terrific job of improving this book. However, the high schools that I've seen really minimize these sections. I hope this person is right. If so, I apologize for not putting more of these problems in this book.

MAX, MIN

The theory of these problems is simplicity itself: To find a maximum or a minimum, take the derivative and set it equal to 0.

In general, there is a picture to be drawn. *Always draw the picture!* Next we have to assign the variable or variables in the problem. Hopefully, by doing enough good examples, you will see how this is done. (I will try my best—you must try your best, which means *don't panic*.) Most of the problems will have two equations in two unknowns. One of these is equal to a number. You will solve for one of the variables and substitute it into the second equation. In the second equation you will take the derivative and set it equal to 0.

Let's start with an easy one.

EXAMPLE 1—

A farmer wishes to make a small rectangular garden with one side against the barn. If he has 200 feet of fence, find the garden of maximum area.

First, we make what I call my crummy little picture. You only have to make the picture good enough so you can understand what you drew.

Second, we assign variables. In this case, this task is easy. The two equations involve the area and the perimeter. $A = xy$. $p = 2x + y$. The trick, if you could call it a trick, is that the barn is one side of the rectangle. The perimeter involves counting y once.

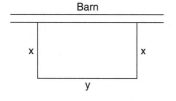

$$p = 200 = 2x + y \quad \text{so} \quad y = 200 - 2x$$

$$A = xy = x(200 - 2x) = 200x - 2x^2$$

$$A' = 200 - 4x = 0 \quad x = 50 \quad y = 200 - 2x = 100$$

The area is x times y, which is 5000 square feet. Sometimes the problem only asks for the dimensions, in which case 50 feet and 100 feet are the answers.

To see whether this is truly a max, we find A″(50). Since A″ = −4, x = 50 is a maximum. We will check only a few times. Sometimes it is too messy. Usually it just takes more time than I feel like taking.

EXAMPLE 2—

A rancher has an 800-square-mile rectangular plot of land he wishes to divide into three equal regions, as pictured below.

   A. Find the dimensions so that the rancher uses the least amount of fencing.

   B. If it costs 4 times as much per mile to fence the outside, find the dimensions needed to minimize the cost.

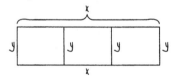

There are two separate problems here. They are done similarly. It is necessary to do them separately, of course.

   A. The area in each case equals length times width. A = xy = 800. So y = 800/x. The fence length f = 2x + 4y = 2x + 4(800/x) = 2x + 3200x^{-1}.

   $$f' = 2 - 3200x^{-2} = 2 - \frac{3200}{x^2} = 0$$

   $$2x^2 = 3200$$

   $$x^2 = 1600$$

   $$x^2 = 40$$

   $$y = 800/x = 20$$

   B. We do not know the cost. So let C equal the cost per mile of the inner fence and 4C the cost per mile of the outer fence. (C is a constant, which we don't know.) The cost of the fence equals the number of fencing miles times the cost per mile.

	Miles of Fence ×	Cost per Mile =	Total Cost
Outside fence	2x + 2y	4C	4C(2x + 2y)
Inside fence	2y	C	2Cy

Total cost $T = 4C(2x + 2y) + 2Cy = 8Cx + 10Cy$. But $A = xy = 800$. So y still equals 800/x. Substitute into T.

$T = 8Cx + 10C(800/x) = 8Cx + 8000Cx^{-1}$

$T' = 8C - 8000Cx^{-2} = 8C - \dfrac{8000C}{x^2} = 0$

$8Cx^2 = 8000C$

$x^2 = 1000$

$x = 10\sqrt{10}$ miles

$y = \dfrac{800}{x} = 800/10\sqrt{10} = 8\sqrt{10}$ miles

EXAMPLE 3—

An open box with a square bottom is to be cut from a piece of cardboard 10 by 10 feet by cutting out the corners and folding the sides up. Find the dimensions that will result in the largest volume.

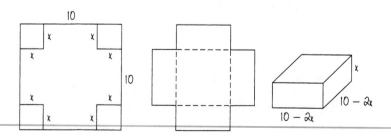

The most difficult part of this problem is the picture, which is given above in three steps. The volume of a box is length times width times height.

$$V = (10 - 2x)(10 - 2x)(x) = 4x^3 - 40x^2 + 100x$$

$$V' = 12x^2 - 80x + 100$$

$$= 4(3x^2 - 20x + 25) = 4(3x-5)(x - 5) = 0$$

$$x = \frac{5}{3} \text{ and } 5$$

We reject 5 since it will give a volume of 0, a minimum.

$$V''(x) = 24x - 80 = 24\left(\frac{5}{3}\right) - 80 = -40$$

indicating a maximum.

The length and width are each $10 - 2x = 10 - 2(5/3) = 20/3$. The box we are looking for is 20/3 by 20/3 by 5/3 feet.

Let's do one more box problem.

EXAMPLE 4—

A box has a square base and no top.

    A. Find the minimum surface area needed if the volume is 4 cubic feet. Let's also do the related problem . . .

    B. Find the maximum volume if the surface area is 12 square feet.

As before, the volume $= (l)(w)(h)$. With a square base, $V = x^2 y$. A box with not top has five surfaces—a square bottom and four sides, all of which have the same dimensions. The surface area $S = x^2 + 4xy$.

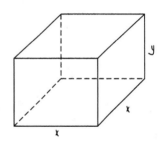

Let's do part A.

$$V = x^2y = 4 \qquad y = \frac{4}{x^2}$$

$$S = x^2 + 4xy = x^2 + 4x\left(\frac{4}{x^2}\right) = x^2 + 16x^{-1}$$

$$S' = 2x - 16x^{-2} = 2x - \frac{16}{x^2} = 0$$

$$2x^3 = 16 \qquad x^3 = 8 \qquad x = 2 \qquad y = \frac{4}{x^2} = 1$$

The box is 2 by 2 by 1 foot. $S = x^2 + 4xy = (2)^2 + 4(2)(1) = 12$ square feet. $S'' = 2 + 32x^{-3}$. $S''(2) = 2 + 32/8 = 6$, which is positive, indicating a minimum.

Now let's do part B. In this case we are maximizing $V = x^2y$ if the surface area $S = x^2 + 4xy = 12$. $y = (12 - x^2)/4x$.

$$V = x^2y = x^2 \frac{12 - x^2}{4x} = \frac{x(12 - x^2)}{4} = 3x - \frac{x^3}{4}$$

$$V' = 3 - \frac{3x^2}{4} = 0 \qquad 3x^2 = 12, x^2 = 4, x = 2$$

$$y = \frac{12 - x^2}{4x} = 1$$

The box is 2 by 2 by 1 foot, and its maximum volume is 4 cubic feet. Of course, we knew this from the first part of the problem. To check this, $V'' = -3x/2$. $V''(2) = -3$, which is negative, indicating a maximum.

EXAMPLE 5—

Find the shortest distance from $y^2 = 2x$ to the point (2, 0).

Although we will draw the picture, it is not truly needed here. Distance means the distance formula. A trick that can always be used is this: Instead of using the distance formula (which involves a square root), use the square of the distance formula (since the distance is a minimum if the square of the distance is the minimum of the square of all the distances).

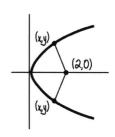

From the picture, we see that two solutions are possible.

Let the square of the distance be $H = (x - 2)^2 + (y - 0)^2 = x^2 - 4x + 4 + y^2$. But on the curve, $y^2 = 2x$. Therefore

$$H = x^2 - 4x + 4 + 2x = x^2 - 2x + 4$$

$$H' = 2x - 2 = 0 \qquad x = 1$$

$$y^2 = 2(1) \qquad \text{so} \qquad y = \pm \sqrt{2}$$

The closest points are $(1, \pm\sqrt{2})$. To find the exact minimum distance, substitute in the distance formula.

The next is a standard-type problem, and many similar kinds are found in all calculus books.

EXAMPLE 6—

Find the dimensions of the largest rectangle (in area) that can be inscribed in the parabola $y = 12 - x^2$, with two vertices of the rectangle on the x axis and two vertices on the curve $y = 12 - x^2$.

Since $y = 12 - x^2$ is symmetric with respect to the y axis, the inscribed rectangle is also symmetric to the y axis. From the picture, the height of the rectangle is y, and the base is $x - (-x) = 2x$.

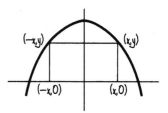

The area $A = 2xy$. Since the point (x, y) is on the curve, $y = 12 - x^2$. Sooooo . . .

$$A = 2xy = 2x(12 - x^2) = 24x - 2x^3$$

$$A' = 24 - 6x^2 = 0 \qquad x = \pm 2$$

$$\qquad\qquad\qquad \text{only } +2 \text{ is used—it is a length}$$

$$y = 12 - x^2 = 8$$

The area is $2xy = 2(2)(8) = 32$.

EXAMPLE 7—

Given a 68-inch string, divide it into two pieces, one a square and the other a rectangle whose length is twice its width, and find the minimum and maximum possible total area of the figures.

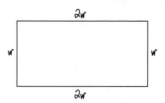

Let one piece be x. The other one will be 68 − x.

If x is the perimeter of the rectangle, then x = 6w. So w = x/6 and l = 2w = x/3. The area of the rectangle is $x^2/18$.

The perimeter of the square is 68 − x. Since p = 4s, s = p/4 = (68 − x)/4. The area of the square is $(68 − x)^2/16$.

The total area A = $x^2/18$ + $(68 − x)^2/16$. dA/dx = x/9 − (68 − x)/8 = 0.

Cross-multiplying after bringing the negative term to the other side, we get

8x = 9(68 − x)   8x = 612 − 9x   17x = 612

Sooo x = 36. The area would be $36^2/18$ + $(68 − 36)^2/16$ = 72 + 64 = 136.

But there are two other possible values for the max and min: If the string is uncut, the whole string may be a rectangle or the whole string might be a square. If the string is a square, s = p/4 = 68/4 = 17. A = s^2 = 289. If the string is a rectangle, w = p/6 = 68/6; l = 2w = 68/3.

$$A = 1 \qquad w = \frac{68}{3} \times \frac{68}{6} = 256.9$$

The largest number is the maximum area, when the figure is a square (289). The smallest number is the minimum area, when the string is cut (36 inches for the perimeter of the rectangle, 32 inches for the rectangle, minimum area = 136).

So far we have limited the problems to the relatively gentle. However, there are some pretty nasty problems in some books. Let's try four of them.

EXAMPLE 8—

Joan's house has a window in the shape of a rectangle surmounted with a semicircle. For a given perimeter p, what are the dimensions of the window if it allows the maximum amount of light?

We first note that maximum light means maximum area. If we are careful and a little clever, the problem is messy but not too bad.

We note that we must maximize the area, which is a rectangle plus one-half of a circle.

$$A = Lw + \frac{\pi r^2}{2} = Lw + \frac{\pi(L/2)^2}{2} = Lw + \frac{\pi L^2}{8}$$

The perimeter consists of three sides of the rectangle plus the circumference of the semicircle.

$$p = L + 2w + \frac{2\pi r}{2} = L + 2w + \pi\frac{L}{2} = \left(1 + \frac{\pi}{2}\right)L + 2w$$

We then solve for w and substitute in the formula for A. We get

$$A = \frac{pL}{2} - \left(\frac{1}{2} + \frac{\pi}{8}\right)L^2$$

Setting dA/dL = 0, we get L = 2p/(4 + π). From the formula for p we get 2w = 2p/(4 + π)!!!!!!!!!!!! The maximum light occurs when the length equals twice the width!!!!?!!!

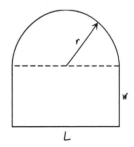

NOTE

In some books, the string is cut into a square and an equilateral triangle, or sometimes—yech!—a square and a circle (quite messy).

There are many problems where we are asked to find the maximum surface area or volume of a figure inscribed in a sphere or other shape. We will look at a messy one.

EXAMPLE 9—

Find the dimensions of the cone of largest volume that can be inscribed in a sphere of radius R.

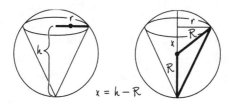

First we note the symmetry of the cone inside the sphere. We note that the variables are r and h, the dimensions of the cone, which are to be found. The only known in the problem is the radius of the sphere R, which is a given number. In the second picture is the way h, r, and R are related, again a right triangle.

$$(h - R)^2 + r^2 = R^2 \qquad \text{or} \qquad r^2 = 2hR - h^2$$

Since the volume of a cone is $(\frac{1}{3}) \pi r^2 h$, it will make a much easier problem to differentiate if we solve for r^2. Thus

$$V = \frac{\pi h(2hR - h^2)}{3} = \frac{\pi(2h^2R - h^3)}{3}$$

$$V' = \left(\frac{\pi}{3}\right)(4hR - 3h^2) = 0$$

$$3h^2 = 4hR \qquad h = \frac{4R}{3}$$

We can substitute into the expression for r^2 and then take the square root to get r.

At this point you might cry, "No more. *No more!!!!*" However, there are two other problems which are in most books. I think you will appreciate seeing them.

EXAMPLE 10—

A man is on an island which is 4 miles from the nearest point on a straight shoreline. He wishes to go to a house which is 12 miles from this nearest point. If he rows at 3 mph and runs at 5 mph, find the shortest time to reach the house.

Insight: If he rowed faster than he ran, he would have rowed straight for the house. If he ran muuuch faster than he rowed, he would row straight for the shore and then run. So we are looking for the point on the shore where he must land. The picture is shown below.

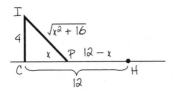

The unknown landing point P is x miles from C, the closest point. We get IP by the Pythagorean theorem. Since CH is 12, PH is 12 − x. We need to set up an equation for time, where time = distance/speed.

	Distance	Speed	Time = $\dfrac{\text{Distance}}{\text{Speed}}$
From I to P (water)	$(x^2 + 16)^{1/2}$	3	$(x^2 + 16)^{1/2}/3$
From P to H (land)	$12 - x$	5	$(12 - x)/5$

Total time $t = \dfrac{(x^2 + 16)^{1/2}}{3} + \dfrac{12}{5} - \dfrac{x}{5}$

$\dfrac{dt}{dx} = \dfrac{x}{3(x^2 + 16)^{1/2}} - \dfrac{1}{5} = 0$

$5x = 3(x^2 + 16)^{1/2} \qquad 25x^2 = 9(x^2 + 16) \qquad 25x^2 = 9x^2 + 144$

$16x^2 = 144 \qquad x^2 = 9 \qquad x = 3$

$t = \dfrac{(3^2 + 16)^{1/2}}{3} + \dfrac{12 - 3}{5} = \dfrac{52}{15} = 3$ hours 28 minutes!

Let us now tackle other kinds of word problems, which should be a little easier.

RELATED RATES

The rates we will talk about refer to how things change with respect to time. Related rates reflect how the rates of two or more variables are connected.

Before we try some problems, we will give an equation and differentiate it with respect to time.

EXAMPLE 11—

$$z = x^2 + y^3$$

x, y, and z are variables, and all are functions of time. What we will do is to differentiate the equation implicitly with respect to time. How does this differ from before?

When we were given $y = (x^2 + 1)^{100}$, we let $u = x^2 + 1$, and

$$\frac{dy}{dx} = \frac{dy}{du} \cdot \frac{du}{dx} \qquad \frac{dy}{du} = 100u^{99} \qquad \frac{du}{dx} = 2x$$

so

$$\frac{dy}{dx} = 100u^{99} \cdot 2x = 200(x^2 + 1)^{99}\,x$$

But if $y = u^{100}$ and we did *not* know what u was (only a function of x),

$$\frac{dy}{dx} = 100u^{99}\,\frac{du}{dx}$$

And if $y = u^{100}$ and u was a function of t (which we did not know),

$$\frac{dy}{dt} = 100u^{99}\,\frac{du}{dt}$$

That's all there is to it.

Let's get back to $z = x^2 + y^3$. Differentiating with respect to time, we get

$$\frac{dz}{dt} = 2x\,\frac{dx}{dt} + 3y^2\,\frac{dy}{dt}$$

How many variables are in this equation? Five!! x, y, dz/dt, dx/dt, and dy/dt. To solve this equation, four of the variables must be given or found from the problem.

Let's try one more.

EXAMPLE 12—

$$x^4 y^7 + \tan x = v^5$$

$$x^4 \cdot 7y^6\,\frac{dy}{dt} + y^7 \cdot 4x^3\,\frac{dx}{dt} + \sec^2 x\,\frac{dx}{dt} = 5v^4\,\frac{dv}{dt}$$

Now we are ready for related-rates problems. In general these problems are easier than max and min problems since the most difficult part of the problem, the original equation, either is given or is much easier to find than before.

EXAMPLE 13—

A girl flying a kite plays out a string at 2 feet per second. The kite moves horizontally at an altitude of 100 feet. If there is no sag in the string, find the rate at which the kite is moving when 260 feet of string has been played out.

We first have to imagine a right triangle with the vertical distance, the horizontal distance, and the string. We then must determine which are the constants and which are the variables. The only constant is that the kite is always 100 feet high. (I wonder how they keep the kite exactly 100 feet high?)

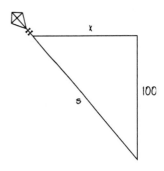

Although we are given the length of the string, it is only at a particular instant. The string length s and horizontal length x are changing.

The equation is $x^2 + 100^2 = s^2$. Differentiating with respect to time, we get

$$2x \frac{dx}{dt} = 2s \frac{ds}{dt} \quad \text{or} \quad x \frac{dx}{dt} = s \frac{ds}{dt}$$

ds/dt is 2. Since s = 260, the Pythagorean theorem tells us that x = 240 feet. (You should have memorized the common Pythagorean triples. This is a 5-12-13 right triangle, or more precisely, a 10-24-26 right triangle.) Substituting in

$$x \frac{dx}{dt} = s \frac{ds}{dt}$$

we get

$$(240) \frac{dx}{dt} = (260)(2)$$

$$\frac{dx}{dt} = \frac{13}{6} \text{ feet per second}$$

EXAMPLE 14—

Sand is leaking from a bag and is forming a cone in which the radius is 6 times as large as the height. Find the rate at which the volume is increasing when the radius is 3 inches and the height is increasing at 2 inches per minute.

We know that the volume of the cone is $V = \frac{1}{3}\pi r^2 h$. There are two variables, but we also know that r = 6h. We can write V in terms of one unknown, but which unknown? Since dh/dt = 2 is given, let us write everything in terms of h.

$$V = \frac{1}{3}\pi(6h)^2h = 12\pi h^3$$

$$\frac{dV}{dt} = 36\pi h^2\left(\frac{dh}{dt}\right) \quad r = 6h \quad \text{and} \quad r = 3 \quad h = \frac{1}{2}$$

$$\frac{dV}{dt} = 36\pi\left(\frac{1}{2}\right)^2(2) = 18\pi \text{ cubic inches per minute}$$

EXAMPLE 15—

When a gas is compressed adiabatically (with no gain or loss of heat), it satisfies the formula $PV^{1.4} = k$; k is a constant. Find the rate at which the pressure is changing if the pressure P is 560 pounds per square inch, the volume is 70 cubic inches, and the volume is increasing at 2 cubic inches per minute.

$PV^{1.4} = k$. Differentiating with respect to time, we get $P(1.4V^{0.4} dV/dt) + V^{1.4} dP/dt = 0$. It looks messy at this point, but fear not.

$$\frac{dP}{dt} = -\frac{1.4PV^{0.4}(dV/dt)}{V^{1.4}} = \frac{-1.4P(dV/dt)}{V}$$

Substituting, we get

$$\frac{dP}{dt} = \frac{-(1.4)(560)(2)}{70} = -22.4 \text{ pounds per square inch per minute}$$

EXAMPLE 16—

At 7 a.m. a plane flies over the city going east at 600 mph. At 9 a.m. a plane flies over the city going north at 350 mph. At what rate were the planes separating at 11 a.m.?

The key is to first make a drawing at the middle time, 9 a.m. The first plane is going east at 600 mph × 2 hours = 1200 miles east of the city and keeps on going east. The distance is 1200 + x. At 9 a.m., the

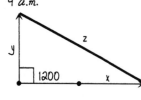

9 a.m.

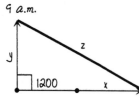

11 a.m.

z = 2500

other plane is over the city and is going north, and distance is $0 + y = y$. Let z be the distance between the planes.

$$z^2 = (x + 1200)^2 + y^2$$

Find dz/dt.

$$2z \frac{dz}{dt} = 2(x + 1200) \frac{dx}{dt} + 2y \frac{dy}{dt}$$

The separating speed is

$$\frac{dz}{dt} = \frac{(x + 1200)\,dx/dt + y\,dy/dt}{z}$$

dx/dt = 600; 9 a.m. to 11 a.m. = 2 hours × 600 = 1200 = x.

dy/dt = 350; time = 2 hours × 350 = 700 = y.
Sticking numbers in the 11 a.m. picture and using the good old Pythagorus theorem, we get z = 2500 (7-24-25 right triangle).

$$\frac{dz}{dt} = \frac{(1200 + 1200)(600) + (700)(350)}{2500}$$

$$= 674 \text{ mph} = \text{separating speed}$$

Finally, sometimes you need two different related-rates equations to solve the problem.

EXAMPLE 17—

Find the rate at which the volume changes with respect to time when the change of surface area with respect to time is 600 square inches per hour and the volume is 1000 cubic inches.

Now $V = x^3$. So dV/dt = $3x^2$ dx/dt. We need to know both x and dx/dt, but we are given neither directly. However, $V = x^3 = 1000$. Sooo x = 10 inches. We are given dS/dt. So the surface area of a cube $S = 6x^2$.

$$\frac{ds}{dt} = 12x \frac{dx}{dt} \qquad \frac{dS}{dt} = 600 \qquad x = 10$$

$$\frac{dx}{dt} = 5 \text{ inches per hour}$$

Finally . . .

$$\frac{dV}{dt} = 3x^2 \frac{dx}{dt} = 3(10)^2(5) = 1500 \text{ cubic inches per hour}$$

THE GRAVITY OF THE SITUATION

The last type of word problem we will deal with involves throwing an object into the air. There are three variables that determine how high an object goes. The first variable is gravity. On this planet, gravity is -32 feet per second squared (or -9.8 meters per second squared); the minus sign indicates down. The second variable is the initial velocity (positive if the object is thrown upward, negative if the object is thrown down, and 0 if the object is dropped). The third variable is the initial height (it goes higher if I throw something from the Empire State Building than from the ground). Interestingly, the weight has no bearing. If I give a rock or a refrigerator the same initial velocity (which I can't, of course . . . or can I?), both will go just as high. Friction, winds, etc. are not included.

The following symbols are usually used:

v_0: initial velocity—the velocity at time $t = 0$

y_0: initial height—the height at time $t = 0$

The initial height is either zero (the ground) or positive. It is very difficult to throw something upward from under the ground. The acceleration is dv/dt. So the velocity is the antiderivative of the acceleration.

The velocity is dy/dt. The height y is the antiderivative of the velocity.

EXAMPLE 18—

On the planet Calculi, gravity is 20 feet per second squared. A ball is thrown upward—initial velocity of 40 feet per second, initial height of 50 feet.

   A. Write the equation for the height.

   B. Find the ball's maximum height.

   C. Find the velocity when the ball hits the ground.

   Now, we proceed as follows:

   A. $v' = a = -20$ (gravity is down). Take the antiderivative—

   $v = -20t + C_1$. $v(0) = 40$, sooooo

   $y' = v = -20t + 40$

   $y = -10t^2 + 40t + C_2$

   $y(0) = 50$, sooooo the equation for the height is

   $y = -10t^2 + 40t + 50$

   B. Throw something lightish into the air straight up. You see that the object stops at its maximum height. This means $v = -20t + 40 = 0$. $t = 2$. Substituting into the equation for the height y, we get $y = -10(2)^2 + 40(2) + 50 = 90$ feet, its maximum height.

   C. The ground, as noted before, is $y = 0$. So

   $0 = -10t^2 + 40t + 50 = -10(t^2 - 4t - 5)$
   $ = -10(t - 5)(t + 1)$

   Rejecting $= -1$ (negative time occurs before the object is thrown upward), we get $t = 5$. Substituting $t = 5$ into the velocity equation, we get

   $v = -20(5) + 40 = -60$ feet per second

   where the minus sign indicates that the object is going down when it hits the ground.

NOTE

Since we live on the planet Earth, it would be a good idea to memorize the formulas for throwing an object upward: $y = -16t^2 + v_0t + y_0$ in feet and $y = -4.9t^2 + v_0t + y_0$ in meters.

CHAPTER 5

INTEGRAL APPLICATIONS

AREAS

We would like to explore some applications of the integral. The first is the area between two curves. Suppose we have two functions f(x) and g(x), where f(x) is always greater than or equal to g(x). Its picture might look like this:

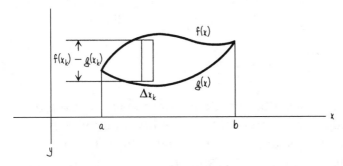

We are going to add these rectangles. One rectangle is represented by the height times the base = $[f(x_k) - g(x_k)]\,\Delta x_k$. If we add them and take the limits properly, we get

$$\int_a^b [f(x) - g(x)]\,dx$$

EXAMPLE 1

Find the area between $y = x + 1$ and $y = 3 - x^2$.

We first draw the curves to see which is the top curve and which is the bottom curve. We next find the limits of integration, the left- and rightmost x values, by setting the curves equal to each other to find the points where they meet.

$x + 1 = 3 - x^2$, $x^2 + x - 2 = 0$, $(x + 2)(x - 1) = 0$.

So $x = -2$ and 1. We then set up the integral of top-curve-minus-bottom-curve dx from -2 to 1.

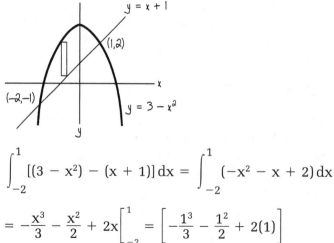

$$\int_{-2}^{1} [(3 - x^2) - (x + 1)]\, dx = \int_{-2}^{1} (-x^2 - x + 2)\, dx$$

$$= -\frac{x^3}{3} - \frac{x^2}{2} + 2x \left.\right]_{-2}^{1} = \left[-\frac{1^3}{3} - \frac{1^2}{2} + 2(1) \right]$$

$$- \left[-\frac{(-2)^3}{3} - \frac{(-2)^2}{2} + 2(-2) \right] = \frac{9}{2}$$

Sometimes it becomes advantageous to set up the rectangles horizontally. We have h(y) and j(y), where j(y) is always larger than or equal to h(y). The picture would look like the one shown at left.

The area of each rectangle would again be the base times the height $= [j(y_k) - h(y_k)]\, \Delta y_k$. If we again properly take the limits, we get

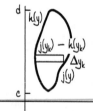

$$\int_{c}^{d} [j(y) - h(y)]\, dy$$

EXAMPLE 2—

Find the area between the curves $x = y^2$ and $x = 3y + 4$.

If we try to make the rectangles as before, the top curve is the parabola, which is OK; but if you look very closely, there are two parts of the bottom curve, a little part of the bottom of the parabola and the straight line. We would have to split the region and set up two integrals to find the area. However, the right curve is always the straight line and the left curve is always the parabola. We take the integral of right-curve-minus-left-curve dy. To find the lowest and highest y values, we again set the curves equal to each other. $y^2 = 3y + 4$, $y^2 - 3y - 4 = 0$, $(y + 1)(y - 4) = 0$. $y = -1$ and $y = 4$.

$$\int_{-1}^{4} [(3y + 4) - y^2]\, dy = \frac{3y^2}{2} + 4y - \frac{y^3}{3} \Big|_{-1}^{4}$$

$$= \left[\frac{3(4)^2}{2} + 4(4) - \frac{4^3}{3}\right] - \left[\frac{3(-1)^2}{2} + 4(-1) - \frac{(-1)^3}{3}\right] = \frac{125}{6}$$

Sometimes splitting the region cannot be avoided. We will do a problem that must be split no matter which way you draw the rectangles. We will do it in two ways. With experience you may be able to tell which way is better *before* you begin the problem.

EXAMPLE 3—

Find the area of the region bounded by $y = x$ to the point $(2, 2)$, then the curve $y = 6 - x^2$ from $(2, 2)$ to $(4, -10)$, and then $y = -5x/2$ from $(4, -10)$ to $(0, 0)$ where it meets the curve $y = x$.

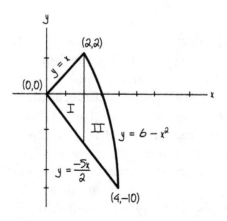

REGION I

$$\int_0^2 \left[x - \left(\frac{-5x}{2} \right) \right] dx = \int_0^2 \frac{7x}{2}\, dx = \frac{7x^2}{4} \Big[_0^2 = \frac{7(2)^2}{4} - 0 = 7$$

REGION II

$$\int_2^4 \left[(6 - x^2) - \left(-\frac{5x}{2} \right) \right] dx = 6x - \frac{x^3}{3} + \frac{5x^2}{4} \Big[_2^4$$

$$= 6(4) - \frac{4^3}{3} + \frac{5(4)^2}{4} - \left[6(2) - \frac{2^3}{3} + \frac{5(2)^2}{4} \right] = 8\frac{1}{3}$$

Total area is $7 + 8\frac{1}{3} = 15\frac{1}{3}$.

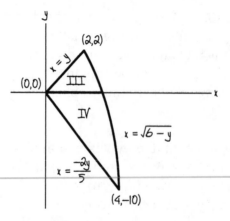

REGION III

$$\int_0^2 [(6 - y)^{1/2} - y]\, dy = -\frac{2}{3}(6 - y)^{3/2} - \frac{y^2}{2} \Big|_0^2$$

$$= \left[-\frac{2}{3}(6 - 2)^{3/2} - \frac{2^2}{2}\right] - \left[-\frac{2}{3}(6 - 0)^{3/2} - 0\right]$$

$$= \frac{2}{3}\, 6^{3/2} - \frac{22}{3}$$

REGION IV

$$\int_{-10}^0 \left[(6 - y)^{1/2} - \left(-\frac{2}{5}\right)y\right] dy = -\frac{2}{3}(6 - y)^{3/2} + \frac{y^2}{5} \Big|_{-10}^0$$

$$= \left[-\frac{2}{3}(6 - 0)^{3/2} + 0\right] - \left[-\frac{2}{3}[6 - (-10)]^{3/2} + \frac{(-10)^2}{5}\right]$$

$$= \frac{68}{3} - \frac{2}{3}\, 6^{3/2}$$

Total area $= \frac{2}{3}\, 6^{3/2} - \frac{22}{3} + \frac{68}{3} - \frac{2}{3}\, 6^{3/2} = \frac{46}{3} = 15\frac{1}{3}$

The two answers agree. The area is $15\frac{1}{3}$.

The last example for us is a region where the curves cross each other (the top curve becomes the bottom and the bottom becomes the top). This is another reason you must draw the region.

EXAMPLE 4—

Find the area of the region between $y = x$ and $y = x^3 - 3x$.

To find the limits, we set $x^3 - 3x = x$. $x^3 - 4x = 0$. $x(x + 2)(x - 2) = 0$. So $x = -2$, 0, and 2. In region I, the top curve is the cubic. So we get $[(x^3 - 3x) - (x)]\, dx$ from -2 to 0. In region II, we have the straight line as the top curve. We get $[x - (x^3 - 3x)]\, dx$ from 0 to 2.

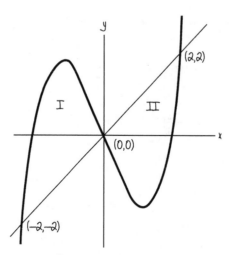

REGION I

$$\int_{-2}^{0} (x^3 - 4x)\,dx = \frac{x^4}{4} - 2x^2 \Big|_{-2}^{0} = 0 \left[\frac{(-2)^4}{4} - 2(-2)^2\right] = 4$$

REGION II

$$\int_{0}^{2} (4x - x^3)\,dx = 2x^2 - \frac{x^4}{4} \Big|_{0}^{2} = \left[2(2)^2 - \frac{2^4}{4}\right] - 0 = 4$$

The total area is 4 + 4 = 8.

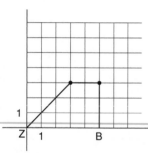

NOTE

If we had noticed that the two regions have the same area, we could have found the area of either one and doubled its value.

Examples 1 through 4 of this chapter are the standard way that college courses ask integral questions. It is the way your teacher will ask you, at least some of the time. However, the AP exam usually asks the questions differently.

EXAMPLE 5—

Find $\int_{Z}^{B} f(x)\,dx$. It involves knowing what the area is. If you do, you can do as little as count the boxes. The area is 4.5 + 6 = 10.5.

EXAMPLE 6—

Given: the integral $\int_z^x (fx)\,dx$. For what value of x do we get the maximum area?

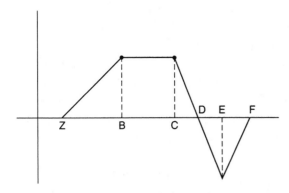

1. B   2. C   3. D   4. E   5. F

The answer is D, or x = 4, since if you go beyond D, you get "negative" area.

As you can see, the AP exam asks thought questions. It has a minimum amount of arithmetic and algebraic skills.

Let's try one more.

EXAMPLE 7—

Given: the region bounded by the x axis, the y axis, and the curve $y = x^2 - 2kx$ in the fourth quadrant. If the area of the region is 36, find k. You must draw the picture. It would look like something like this:

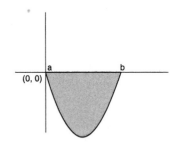

It is clear the area is $\int_a^b [0 - (x^2 - 2kx)]\,dx = 36$.

The left limit a = 0. To find b, we let $x^2 - 2kx = 0$. So x = 2k also.

$$\int_0^{2k} (2kx - x^2)\,dx = kx^2 - \frac{x^3}{3}\Big|_0^{2k} = 4k^3 - \frac{8k^3}{3} = \frac{4k^3}{3} = 36$$

Solving, we get k = 3.

VOLUMES OF ROTATIONS

The next topic involves finding volumes of rotations. This is very visual. If you see the picture, the volume is easy. If not, this topic is very hard.

Imagine a perfectly formed apple with a line through the middle from top to bottom. We can find the volume in two different ways. One way is by making slices perpendicular to the line (axis). (We will do the other way later with an onion.) Each slice is a disc, a thin cylinder. Its volume is $\pi r^2 h$, where h is very small. If we add all the discs, taking the limits properly, we get the volume.

We will take the same region in six different problems, rotating this region differently six times and getting six different volumes.

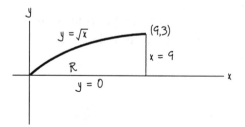

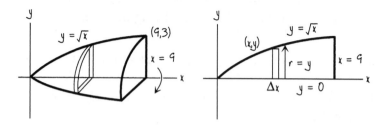

EXAMPLE 8

Find the volume if the region R is rotated about the x axis.

The volume of each disc is $\pi r^2 h$. $h = \Delta x$. $r = y$.
So $r^2 = y^2 = x$; x goes from 0 to 9.

$$V = \int_0^9 \pi r^2 \, dh = \pi \int_{x=0}^9 x \, dx = \frac{\pi x^2}{2} \Big[_0^9 = \frac{81\pi}{2}$$

The integrals are almost always easy. Once you understand the picture, all will be easy. But it takes most
people time to study the pictures.

Let's get back to the apple. Suppose we core the apple.
When we take slices perpendicular to the axis, we get
rings. The area of a ring is the area of the outside minus
the area of the inside. The volume of each disc is

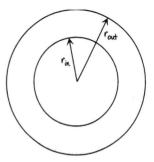

$$(\pi r_{out}^2 - \pi r_{in}^2)h$$

Again, h is small.

EXAMPLE 9—

Find the volume if our region is rotated about the y axis.

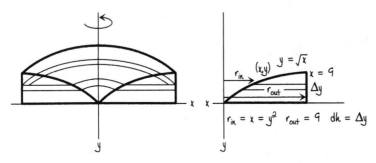

As you rotate this region, there is a hole. The outside
radius is always 9 and the inside radius is always the
x value. But $x = y^2$. $r^2 = x^2 = y^4$.

$$V = \int_{y=0}^3 (\pi r_{out}^2 - \pi r_{in}^2) \, dh = \pi \int_0^3 (9^2 - y^4) \, dy$$

$$= \pi \left[81y - \frac{y^5}{5} \right]_0^3 = \pi \left[81(3) - \frac{3^5}{5} \right] = \frac{972}{5} \pi$$

EXAMPLE 10—

Find the volume of our glorious region if it is rotated about the line x = 9.

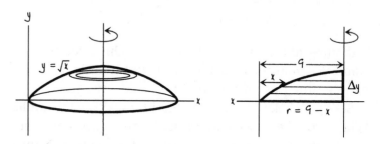

Notice that when we rotate the region about x = 9, there is no hole. $V_{sect} = \pi r^2 h$. $r = 9 - x = 9 - y^2$. $r^2 = 81 - 18y^2 + y^4$. $h = \Delta y$.

$$V = \pi \int_0^3 (81 - 18y^2 + y^4)\,dy = 81y - 6y^3 + \frac{y^5}{5}\Big|_0^3 \pi$$

$$= \pi\left[81(3) - 6(3)^3 + \frac{3^5}{5} - 0\right] = \frac{648\pi}{5}$$

EXAMPLE 11—

Find the volume if the same region R is rotated about the line x = −1.

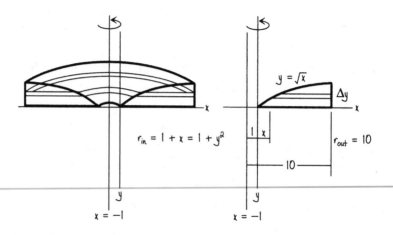

$V_{sec} = \pi(r_{out}^2 - r_{in}^2)h.\ r_{out} = 9 + 1 = 10.\ r_{in} = 1 + x = 1 + y^2.\ r^2 = 1 + 2y^2 + y^4.\ h = \Delta y.$

$$V = \int_0^3 (\pi r_{out}^2 - \pi r_{in}^2)\, dy = \pi \int_0^3 [10^2 - (1 + 2y^2 + y^4)]\, dy$$

$$= \pi \int_0^3 (99 - 2y^2 - y^4)\, dy = 99y - \frac{2y^3}{3} - \frac{y^5}{5}\ \Big[_0^3\ \pi$$

$$= \pi \left[99(3) - \frac{2(3)^3}{3} - \frac{3^5}{5} - 0 \right] = \frac{1152\pi}{5}$$

EXAMPLE 12—

Find the volume if our beloved region is rotated about $y = 5$.

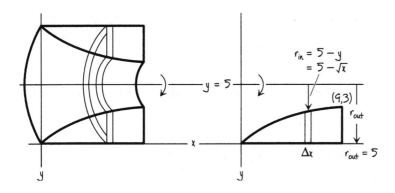

$V_{sec} = \pi(r_{out}^2 - r_{in}^2)h.\ r_{out} = 5.\ r_{in}\ 5 - y = 5 - x^{1/2}.\ r_{in}^2 = 25 - 10x^{1/2} + x.\ h = \Delta x.$

$$V = \int_0^9 (\pi r_{out}^2 - \pi r_{in}^2)\, dx = \pi \int_0^9 [25 - (25 - 10x^{1/2} + x)]\, dx$$

$$= x \int_0^9 (10x^{1/2} - x)\, dx = \frac{20}{3} x^{3/2} - \frac{x^2}{2}\ \Big[_0^9\ \pi$$

$$= \pi \left(\frac{20}{3} \cdot 9^{3/2} - \frac{9^2}{2} - 0 \right) = 139.5\pi$$

EXAMPLE 13—

And for our final attraction we will take the same region and rotate it about the line $y = -2$.

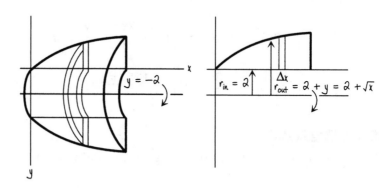

$V_{sect} = (r_{out}^2 - r_{in}^2)h.$ $r_{in} = 2$, $r_{out} = 2 + y = 2 + x^{1/2}$. $r_{out}^2 = 4 + 4x^{1/2} + x$.

$$V = \int_0^9 (\pi r_{out}^2 - \pi r_{in}^2)\,dx = \pi \int_0^9 [(4 + 4x^{1/2} + x) - 4]\,dx$$

$$= \pi \int_0^9 (4x^{1/2} + x)\,dx = \frac{8}{3}x^{3/2} + \frac{x^2}{2} \Big|_0^9 \, \pi$$

$$= \pi\left(\frac{8}{3} \cdot 9^{3/2} + \frac{9^2}{2}\right) = \frac{225\pi}{2}$$

The next kinds of volume we will consider are rotations again, but we will do it in a different way. Think about an onion with each layer a whole piece. We will add layer by layer until we get a volume. We will add *cylindrical shells*, tall cylindrical shells. We will see one of these shells, a general picture, and two examples.

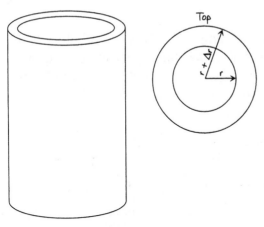

$$V_{shell} = \pi r_{out}^2 h - \pi r_{in}^2 h$$
$$= \pi h(r_{out}^2 - r_{in}^2)$$
$$= \pi h(r_{out} + r_{in})(r_{out} - r_{in})$$
$$= 2\pi h\left(\frac{r_{out} + r_{in}}{2}\right)(\Delta r)$$

$$= 2\pi \cdot \text{height} \cdot \text{average radius} \cdot \text{thickness}$$

This is an example of a rotation about the y axis of the region bounded by $y = f(x)$, $y = 0$, $x = a$, $x = b$. *Notice that the axis of the cylinder is the axis which the curve is rotated about.*

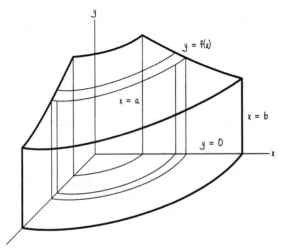

EXAMPLE 14

Let us try to take the same region R we did before and rotate it about the y axis. We take the same region for two reasons: (1) we do not have to worry about different curves, and (2) we would like to show that the answers are the same.

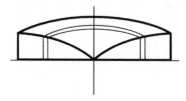 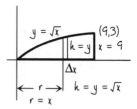

$$V = \int 2\pi \cdot \text{average radius} \cdot \text{average height} \cdot \text{thickness}$$

$$= 2\pi \int_0^9 x \cdot x^{1/2}\, dx = 2\pi \int_0^9 x^{3/2}\, dx = \frac{4}{5}x^{5/2}\Big[_0^9 \pi$$

$$= \frac{4\pi}{5}(9^{5/2} - 0) = \frac{972\pi}{5}$$

EXAMPLE 15

We will rotate our region one last time about the x axis.

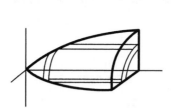 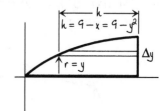

$$V = 2\pi \int_{y=0}^3 \text{average radius} \cdot \text{average height} \cdot \text{thickness}$$

$$= 2\pi \int_0^3 y(9 - y^2)\, dy = 2\pi \int_0^3 (9y - y^3)\, dy$$

$$= \frac{9y^2}{2} - \frac{y^4}{4}\Big[_0^3 2\pi = 2\pi\left[\frac{9(3)^2}{2} - \frac{3^4}{2}\right] = \frac{81\pi}{2}$$

VOLUMES BY SECTION

The last part of this chapter covers volumes by section. The sections are not circles but are other shapes. Two examples are given.

EXAMPLE 16—

Find the volume of the following figure: base bounded by $y = \frac{1}{4} x^2$ and $y = 4$ in the xy plane. Sections perpendicular to the y axis are rectangles with heights of one-third the base.

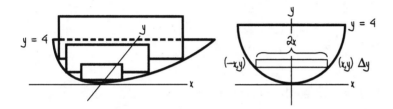

$$A = bh = b\left(\frac{b}{3}\right) = \frac{b^2}{3} \qquad b = x - (-x) = 2x$$

$$y = \frac{x^2}{4} \qquad x^2 = 4y \qquad x = 2y^{1/2} \qquad 2x = 4y^{1/2} = b$$

$$A = \frac{(4y^{1/2})^2}{3} = \frac{16y}{3}$$

Volume of each section $= (16y/3) \, \Delta y$

$$V = \int_0^4 \frac{16y}{3} \, dy = \frac{8y^2}{3}\left[\begin{matrix}4\\0\end{matrix}\right. \frac{8(4)^2}{3} - \frac{8(0)^2}{3} = \frac{128}{3}$$

EXAMPLE 17—

Find the volume of the following figure. Again the base is bounded by $y = \frac{1}{4}x^2$ and $y = 4$ in the xy plane. Sections perpendicular to the x axis are right isosceles triangles with one of the legs on the xy plane.

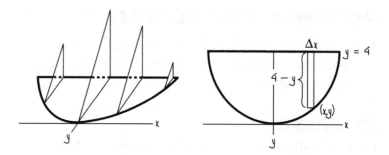

$A = \frac{1}{2}bh = \frac{1}{2}b^2$ (*isosceles* means two equal legs)

$b = 4 - y = 4 - \frac{x^2}{4}$

$A = \frac{1}{2}\left(4 - \frac{x^2}{4}\right)^2 = 8 - x^2 + \frac{x^4}{32}$

Volume of each section $= \left(8 - x^2 + \frac{x^4}{32}\right) \Delta x$

$$V = \int_{-2}^{2}\left(8 - x^2 + \frac{x^4}{32}\right)dx = 8x - \frac{x^3}{3} + \frac{x^5}{160} \Bigg|_{-2}^{2}$$

$$= \left[8(2) - \frac{2^3}{3} + \frac{2^5}{32}\right] - \left[8(-2) - \frac{(-2)^3}{3} + \frac{(-2)^5}{32}\right] = \frac{406}{15}$$

If you draw your pictures very carefully, this figure is symmetric with respect to the y axis. You can make the problem easier by doubling the integral from 0 to 2. That is, y axis symmetry means

$$\int_{-a}^{a} f(x)\,dx = 2\int_{0}^{a} f(x)\,dx$$

Following the section on volumes of rotations, some of the students wondered if you always have a π when you do volumes by sections. These two additional examples deal with that question.

EXAMPLE 18—

Base $x^2 + y^2 = 9$. Sections perpendicular to the x axis
are equilateral triangles. Find the volume.

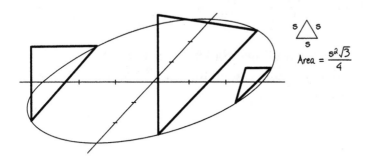

$$\int_{-3}^{3} A(x)\,dx \quad A = \frac{s^2\sqrt{3}}{4} = \frac{(2y)^2\sqrt{3}}{4} = \sqrt{3}y^2 = \sqrt{3}(9 - x^2)$$

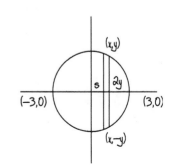

$$\sqrt{3}\int_{-3}^{3} (9 - x^2)\,dx = \sqrt{3}\left[9x - \frac{x^3}{3}\right]_{-3}^{3} = \sqrt{3}[18 - (-18)]$$

$$= 36\sqrt{3}$$

Even though the base is a circle, we are adding triangu-
lar slices. Since the area of a triangle does not involve π,
neither does our volume.

EXAMPLE 19—

Base is bounded by $y = x^{1/2}$, $x = 9$, x axis. Sections
perpendicular to the x axis are semicircles. Find the
volume.

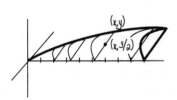   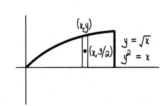

$$\int_0^9 A(x)\,dx \qquad A = \frac{1}{2}\pi r^2 = \frac{1}{2}\pi\left(\frac{y}{2}\right)^2 = \frac{\pi}{8}y^2 = \frac{\pi x}{8}$$

$$\frac{\pi}{8}\int_0^9 x\,dx = \frac{\pi x^2}{16}\Big[_0^9 = \frac{81\pi}{16}$$

Since we are adding circular slices (well, to be absolutely accurate, semicircular slices), the volume does have a π in it.

CHAPTER 6

ODDS AND ENDS

This is the chapter that covers everything we don't
want to put in any of the other chapters.

DIFFERENTIALS

The first topic is differentials. One of the wonderful
things calculus does is to give techniques for approximations. Since we live in an imperfect world (really?),
approximations are very necessary (exact answers are
nicest, of course, but are not usually necessary and
sometimes not attainable). Later we will get more
sophisticated means of approximation. This method is
usually the first.

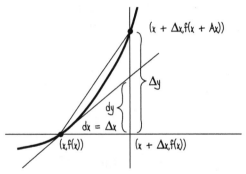

DEFINITION

Differential—dy $= f'(x)\,dx$. [y $= f(x)$.]

Δy is the actual change; dy is the approximate change. If Δx is small when compared to x, then dy is a good approximation for Δy. In the picture Δx is large. Otherwise you couldn't see the picture.

EXAMPLE 1—

Find the differential dy if $y = f(x) = x^3 + x^2$.

$dy = f'(x)\,dx = (3x^2 + 2x)\,dx$

EXAMPLE 2—

A 9-inch steel cube is heated so that each side is increased by $\frac{1}{100}$ inch. Find the actual change and the approximate change in the volume.

$$V = x^3 \qquad \Delta V = (9.01)^3 - 9^3 = 731.432701 - 729$$

$$= 2.432701$$

$$dV = V'(x)\,dx = 3x^2\,dx = 3(9)^2\,(0.01) = 2.43$$

Notice the difference between ΔV and dV, $2.432701 - 2.43 = 0.002701$, is very small, especially when compared to a volume of 729 cubic inches, since $dx = 0.01$ is small when compared to $x = 9$.

EXAMPLE 3—

Approximate $\sqrt[3]{63.3}$ by using differentials.

We first locate the closest number that is an exact cube: $x = 64$. $dx = 63.3 - 64 = -0.7$. Computing the differential at $x = 64$, $y = x^{1/3}$ gives

$$dy = \frac{dx}{3x^{2/3}} = \frac{-0.7}{3(64)^{2/3}} = \frac{-0.7}{3(16)} = \frac{-7}{480}$$

So our approximation is

$$\sqrt[3]{63.3} = \sqrt[3]{64} + dy = 4 - \frac{7}{480} = 3\frac{473}{480}$$

MEAN VALUE THEOREM

The mean value theorem (MVT) is a very important theorem in differential calculus that periodically pops up to give us more useful information.

The mean value theorem: Let $f(x)$ be continuous on $[(a, b)]$. Let $f'(x)$ exist on (a, b). There exists a point c between a and b such that $f'(c) = [f(b) - f(a)]/(b - a)$.

Let us translate. There are no breaks between a and b, including both ends. The derivative exists except at possibly the left and right ends. Otherwise it is smooth. $f'(c)$ is the slope at c. What the heck is the rest? If $x = a$, the y value is $f(a)$. If $x = b$, the y value is $f(b)$. The two endpoints are $(a, f(a))$ and $(b, f(b))$. The slope of the line joining the two endpoints is $[f(b) - f(a)]/(b - a)$.

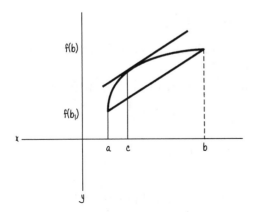

The theorem says there is at least one point between a and b where the curve has the same slope as the line joining the endpoints of the curve.

We will discuss this theorem fully and give examples. At the end, we will prove the theorem.

QUESTION I
Can there be more than one such point?

The answer is yes, but one and only one is guaranteed by the MVT.

QUESTION 2

Can there be such a point if the continuity or differentiability condition does not hold? The answer is yes, but there also may not be such a point. In both cases illustrated below, there is no point on the curve where the slope is the same as the slope of the line joining the endpoints of the curve.

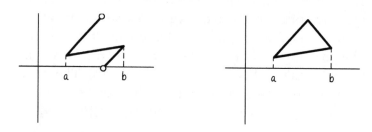

EXAMPLE 4

Find the point c in the MVT or explain why it does not exist if $f(x) = x^2 + 3x + 5$ on $[1, 4]$.

Polynomials are always continuous and differentiable everywhere. The MVT holds. $f'(x) = 2x + 3$. $f'(c) = 2c + 3$. Therefore

$$2c + 3 = \frac{f(4) - f(1)}{4 - 1} = \frac{33 - 9}{3} \qquad 2c + 3 = 8 \qquad c = 2.5$$

Notice that 2.5 is between 1 and 4. Let's try others.

EXAMPLE 5

Same as above if

$$f(x) = \frac{4}{x - 3} \qquad \text{on } [1, 4]$$

The MVT does not hold since the function is not continuous at $x = 3$.

EXAMPLE 6

Same as above except $f(x) = (x - 4)^{1/3}$ on $[0, 10]$.

$f'(x) = 1/[3(x - 4)^{2/3}]$ and the derivative does not exist at $x = 4$. The MVT again does not hold.

EXAMPLE 7—

Same as above except $f(x) = x^{1/3}$ on $[0, 8]$.

$f'(x) = 1/(3x^{2/3})$. The function is continuous. The derivative exists at every point except $x = 0$. The MVT holds because $f'(x)$ does not have to exist at the endpoints.

$$f'(c) = \frac{1}{3c^{2/3}} = \frac{f(8) - f(0)}{8 - 0} = \frac{2}{8} = \frac{1}{4}$$

$$3c^{2/3} = 4 \quad \text{so} \quad c = \left(\frac{4}{3}\right)^{3/2}$$

NOTE

You alert students might have seen in Example 6 that the MVT does not hold, yet there is a point that satisfies the theorem. This is possible. If the MVT holds, such a point c is guaranteed. If the MVT doesn't hold, such a point is possible but not guaranteed.

To prove the MVT, we will need Rolle's theorem.

ROLLE'S THEOREM

Let $f(x)$ be continuous on $a \leqslant x \leqslant b$. Let $f'(x)$ exist on $a < x < b$. Finally let $f(a) = f(b)$. Then there exists a point c between a and b such that $f'(c) = 0$.

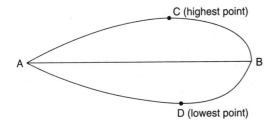

There are three possibilities:

1. $f(x)$ = constant, as in line segment $\overline{AB}$. For any point between a and b, $f'(c) = 0$.

2. If $f(x)$ is not a constant, suppose it goes up as in arc ACB. At the beginning $f'(x) > 0$. At the end $f'(x) < 0$. Since the slope exists at every point in between, at some point c, in this case at C, $f'(c) = 0$.

3. Suppose $f(x)$ goes down as in arc ADB. The same argument holds. The slope is negative at the left end and positive at the right. Since the slope exists at every point in between, there exists a point c, in this case at D, such that $f'(c) = 0$.

Mathematicians want the most general proof possible.

As we have seen, the first two conditions cannot be eliminated; $f(a) = f(b)$ can be eliminated.

Let us restate the MVT and prove it. The proof is very clever.

MEAN VALUE THEOREM

Let $f(x)$ be continuous on [a, b].

Let $f'(x)$ exist on (a, b). Then there exists a point c such that $f'(x) = \dfrac{f(b) - f(a)}{b - a}$. If you look at the picture, the MVT is Rolle's theorem tilted.

We will create a function $F(x)$ that is the difference between $f(x)$ and the line joining the ends of the line segment joining (a, f(a)) and (b, f(b)).

We will show $F(x)$ satisfies Rolle's theorem. The equation of the line segment using point-slope form is the following:

By using (a, f(a)) as the point, $m = \dfrac{y - y_1}{x - x_1}$ becomes

$$\frac{f(b) - f(a)}{b - a} = \frac{y - f(a)}{x - a}.$$

Solving for y, we get $y = \dfrac{f(b) - f(a)}{b - a}(x - a) + f(a)$.

Let $F(x) = f(x) - \left\{ \left[\dfrac{f(b) - f(a)}{b - a}(x - a) \right] + f(a) \right\}$.

Oh, let's redraw the figure.

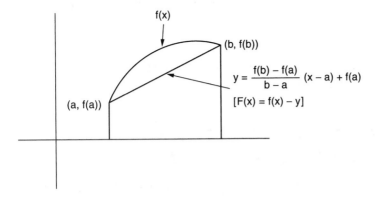

$F(x)$ is continuous on $[a, b]$ since $f(x)$ and lines are continuous.

$F(x)$ is differentiable on (a, b) since $f(x)$ and lines are differentiable.

$F(a) = f(a) - f(a) = 0$ and $F(b) = f(b) - f(b) + f(a) - f(a) = 0$. So $F(a) = F(b)$.

Rolle's theorem holds! So there is a point c between a and b such that $F'(c) = 0$.

$F'(c) = f'(c) - \dfrac{f(b) - f(a)}{b - a} = 0$, since the derivative of $(x - a)$ is 1.

Solving, we get $f'(c) = \dfrac{f(b) - f(a)}{b - a}$.

NOTE

Let's show the derivative of $F(x)$ in detail.

1. The derivative of f(x) is f'(x).

2. The derivative of $\dfrac{f(b) - f(a)}{b - a}$ (x − a) is $\dfrac{f(b) - f(a)}{b - a}$ since $\dfrac{f(b) - f(a)}{b - a}$ is a number, and the derivative of a number times (x − a) is that number [the derivative of (x − a) is 1].

3. The derivative of f(a) [−f(a)] is 0 since −f(a) is a constant.

APPROXIMATIONS, APPROXIMATIONS

In an age in which computers and calculators, even those fun graphing calculators, do so many things, some things cannot be done exactly.

We know that all quadratics can be solved by using the quadratic formula. Similarly, there is a cubic formula and a quartic (fourth-degree) formula that can solve all cubics or quartics (although they are truly ugly and messy). However, in higher mathematics we can prove that most general fifth-degree equations cannot be solved. More simply, an equation such as 2x = cos x cannot be solved exactly. However, we can approximate a solution very closely.

Newton's Method

Suppose we have an equation y = f(x). Let's say it crosses the x axis at x = r. That is, the root f(r) = 0, but it is not exact. If we can find f(a) < 0 and f(b) > 0, then if f(x) is continuous, there is a point r such that f(r) = 0, and r is between a and b. We can use Newton's method.

1. Let x_1 be the first approximation. Draw a tangent line at the point $(x_1, f(x_1))$ until it hits the x axis at x_2, which is usually closer to r than x_1. Continue. . . .

2. Let us give a formula using point slope.

$$y - f(x_1) = f'(x_1)(x - x_1)$$

If $f'(x_1) \neq 0$, the line is not parallel to the x axis, and the line hits the x axis at, let's say, the point $(x_2, 0)$. Substitute the point in step 2.

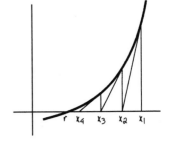

3. $0 - f(x_1) = f'(x_1)(x_2 - x_1)$. We solve for x_2.

4. $x_2 = x_1 - \dfrac{f(x_1)}{f'(x_1)}$

Repeating, we get the general formula

$$x_{n+1} = x_n - \frac{f(x_n)}{f'(x_n)}$$

Let's do an example.

EXAMPLE 8—

Find the root of $f(x) = x^3 - x - 3$ by using Newton's method.

The picture of the graphing, using the fun TI-82 calculator, looks like this:

$$f(x) = x^3 - x - 3$$

Therefore $f'(x) = 3x^2 - 1$.

Newton's formula becomes

$$x_{n+1} = x_n - \frac{f(x_n)}{f'(x_n)}$$

$$= x_n - \frac{x_n^3 - x_n - 3}{3x_n^2 - 1}$$

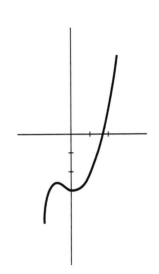

Simplifying, we get

$$x_{n+1} = \frac{2x_n^3 + 3}{3x_n^2 - 1}$$

$f(1) = -3$ and $f(2) = 3$.

Let's take $x_1 = 1.5$ for the first approximation. Probably $x_1 = 1$ or $x_1 = 2$ would work OK.

$$x_2 = \frac{2(1.5)^3 + 3}{3(1.5)^2 - 1} = 1.695652174$$

$$x_3 = \frac{2(1.695652174)^3 + 3}{3(1.695652174)^2 - 1} = 1.672080792$$

Continuuuuuing, we get $x_4 = 1.671700271$, $x_5 = 1.671699882$, $x_6 = 1.671699882$. We have reached the accuracy of our calculator—nine decimal places—probably more accurate than we would ever need.

NOTE 1

Sometimes the method doesn't work. A full study is left to other courses.

NOTE 2

When the method works, it usually works very quickly with great accuracy.

NOTE 3

This topic, as well as the other two approximations, can be studied at many levels. Most appropriately, they occur in either Calc I or Calc II.

Trapezoidal Method

This method, the only one of the three that does not actually require calculus, approximates the area under the curve by trapezoids, by approximating the "top" of the region by a line.

Divide the region into n equal parts. See the trapezoids. They are standing on their heights. The area of a trapezoid is $\frac{1}{2}h(b_1 + b_2)$. All the h's are the same. For the first trapezoid, $b_1 = y_1$ and $b_2 = y_2$. $A = \frac{1}{2}h(y_2 + y_2)$. For the second trapezoid, $b_1 = y_2$ and $b_2 = y_3$. $A = \frac{1}{2}h(y_2 + y_3)$.

Notice that the lower base of the first trapezoid is the upper base of the second trapezoid. Every base is doubled except the first upper base and the last lower base. The formula is $A = \frac{1}{2}h(y_0 + 2y_1 + 2y_2 + \cdots + 2y_{n-1} + y_n)$.

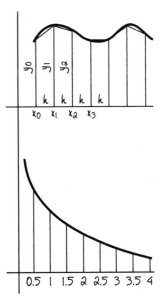

EXAMPLE 9—

Approximate $\int_1^4 \frac{1}{x}\,dx$ by using six equal subdivisions.

The interval is of length $4 - 1 = 3$.
Six equal parts? Each $h = \frac{3}{6} = \frac{1}{2}$.

$x_0 = 1 \qquad y_0 = \frac{1}{1} = 1$

$x_1 = \frac{3}{2} \qquad y_1 = 1\left(\frac{3}{2}\right) = \frac{2}{3}$

$x_2 = 2, y_2 = \frac{1}{2} \qquad x_3 = \frac{5}{2}, y_3 = \frac{2}{5} \qquad y_4 = \frac{1}{3} \qquad y_5 = \frac{2}{7} \qquad y_6 = \frac{1}{4}$

$A = \frac{1}{2}h(y_0 + 2y_1 + 2y_2 + 2y_3 + 2y_4 + 2y_5 + y_6)$

$\quad = \frac{1}{4}\left[1 + 2\left(\frac{2}{3}\right) + 2\left(\frac{1}{2}\right) + 2\left(\frac{2}{5}\right) + 2\left(\frac{1}{3}\right) + 2\left(\frac{2}{7}\right) + \frac{1}{4}\right]$

$\quad = 1.405357143$

Parabolic Method

Another method is to approximate the region by using parabolas on the top. We will isolate one of these regions. As you will see, only if n = even, an even number of intervals, this method will work.

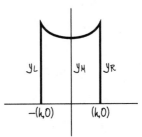

Let the parabola be given by $y = f(x) = ax^2 + bx + c$.
y at the left end is $y_L = f(-h) = ah^2 - bh + c$.
y in the middle is $y_M = f(0) = c$.
y at the right is $y_R = f(h) = ah^2 + bh + c$.
The area of this region is

$$\int_{-h}^{h} (ax^2 + bx + c)\, dx$$

$$= \frac{ax^3}{3} + \frac{bx^2}{2} + cx \Big|_{-h}^{h}$$

$$= \frac{ah^3}{3} + \frac{bh^2}{2} + ch - \left(-\frac{ah^3}{3} + \frac{bh^2}{2} - ch \right)$$

Do something "weird." Factor out h/3 because it works!!

$$= \frac{2ah^3}{3} + 2ch$$

$$= \frac{h}{3} (2ah^2 + 6ch)$$

$$= \frac{h}{3} (2ah^2 + 2ch + 4ch)$$

$$= \frac{h}{3} (y_L + y_R + 4y_M)$$

Just as before, the lower base of the first region is the upper base of the second region. Four times the middle never changes.

The formula issss . . . $A = \frac{1}{3}h(y_0 + 4y_1 + 2y_2 + 4y_3 + \cdots + 4y_{n-1} + y_n)$, n even.

EXAMPLE 10—

Let's do the same example.
Approximate

$$\int_{1}^{4} \frac{1}{x}\, dx$$

$$A = \frac{h}{3}(y_0 + 4y_1 + 2y_2 + 4y_3 + 2y_4 + 4y_5 + y_6)$$

$$A = \frac{1}{6}\left[1 + 4\left(\frac{2}{3}\right) + 2\left(\frac{1}{2}\right) + 4\left(\frac{2}{5}\right) + 2\left(\frac{1}{3}\right) + 4\left(\frac{2}{7}\right) + \frac{1}{4}\right]$$

$$= 1.387698413$$

As some of you know and the rest will find out soon, the exact answer is $\ln 4 = 1.386294361$, approximately.

Work, Work, Work

This topic is usually presented in a physics book and scares everyone to death. If it is done my way, I'm pretty sure it won't bother you again. Work is defined as force times distance. However if the force is a function of distance, theory tells us that work is the integral of $F(x)\,dx$. We give the usual examples: springs and a couple on pumping water over the top.

EXAMPLE 11

It takes a force of 20 pounds to stretch a 7-foot spring to 11 feet. How much work is needed to pull out the spring from 13 to 27 feet?

For a spring force $F = kx$. $F = 20$, and $x = 11 - 7 = 4$, the distance the spring is stretched from its natural length. So $k = 5$ and $F = 5x$.

$$\text{Work} = \int kx\,dx$$

Lower limit = $13 - 7 = 6$, upper limit $27 - 7 = 20$.

$$\text{Work} = \int_6^{20} 5x\,dx = \left(\frac{5x^2}{2}\right)\Big|_6^{20} = \frac{5(20)^2}{2} - \frac{5(6^2)}{2}$$

$$= 1000 - 90 = 910 \text{ foot-pounds}$$

EXAMPLE 12—

How much work is done to pump the water out of a full cylindrical can of radius 10 feet, height 20 feet—if the water is to be pumped over the top?

W = density (weight/volume) × volume × height over which the water is pumped. Each section of water pumped is a thin cylinder of volume $V = \pi(10)^2\,dy$. Height pumped (see the figure) is $20 - y$. The density for water is approximately 62.5 pounds per cubic foot.

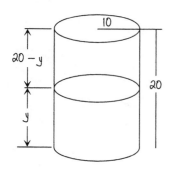

$$\int_0^{20} 62.5\pi(10)^2(20 - y)\,dy = 6250\pi\left(20y - \frac{y^2}{2}\right]_0^{20}$$

$$= 1{,}250{,}000\pi \text{ foot-pounds}$$

NOTE I

If the outlet were 17 feet over the can, we would have $37 - y$.

NOTE 2

If the can were three-fourths full, integral limits would be 0 to 15.

NOTE 3

If we did the same problem with a box, the cross section would be a thin sheet with length and width constant and height dy.

EXAMPLE 13—

How much work is done to pump water out of a cone, of diameter 22 feet and height 10 feet, if the outlet is 7 feet over the top of the cone and the cone is filled 2 feet deep?

First, a slight trick. The diameter is 22, so the radius is 11. Next, note the cross section is again a thin cylinder, but this time the radius changes: $V = \pi r^2 h = \pi x^2\,dy$. We must see a similar triangle. $x/y = 11/10$ or $x = 11y/10$.

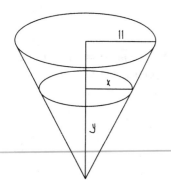

The height of the pipe makes the pumping distance
$17 - y$.

We leave the integration to you.

$$\int_0^2 62.5\pi\left[\left(\frac{11}{10}\right)y\right]^2 (17 - y)\,dy$$

NOTE

A trough is a similar problem except the horizontal
cross section is a rectangular sheet whose width
changes but whose length (length of the trough)
remains the same.

Also note similar triangles, just as in the cone.

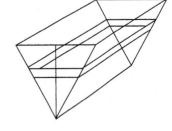

CHAPTER 7

LOGARITHMS

Many of you, at this point in your mathematical journey, have not seen logs for a while. Let's revisit the basics of logarithm be we do the calculus with it.

THE BASIC LAWS OF LOGS

1. Defined, $\log_b x = y$ (log of x to the base b is y) if $b^y = x$; $\log_5 25 = 2$ because $5^2 = 25$.

2. What can the base b be? It can't be negative, such as -2, since $(-2)^{1/2}$ is imaginary. It can't be 0, since 0^n is either equal to 0 if n is positive or undefined if n is 0 or negative. Also b can't be 1 since 1^n always $= 1$. Therefore b can be any positive number except 1.

NOTE

The base can be $2^{1/2}$, but it won't do you any good because there are no $2^{1/2}$ tables. The two most common bases are 10, because we have 10 fingers, and e, a number that occurs a lot in mathematics starting now.

   A. e equals approximately 2.7.

   B. What is e more exactly? On a calculator press 1, inv, ln.

   C. $\log = \log_{10}$.

   D. $\ln = \log_e$ (ln is the natural logarithm).

3. A log y is an exponent, and exponents can be positive, negative, and zero. The range is all real numbers.

4. Since the base is positive, whether the exponent is positive, zero, or negative, the answer is positive. The domain, therefore, is positive numbers.

NOTE

To avoid getting too technical, most books write log $|x|$, thereby excluding only x = 0.

   5. $\log_b x + \log_b y = \log_b xy$; $\log 2 + \log 3 = \log 6$.

   6. $\log_b x - \log_b y = \log_b (x/y)$; $\log 7 - \log 3 = \log (7/3)$.

   7. $\log_b x^p = p \log_b x$; $\ln 6^7 = 7 \ln 6$ is OK.

NOTE

Laws 5, 6, and 7 are most important. If you can simplify by using these laws, about half the battle (the easy half) is done.

EXAMPLE 1—

Write the following as simpler logs with no exponents:

$$\ln\left(\frac{a^4 b^5}{c^6 \sqrt{d}}\right)$$

$4 \ln a + 5 \ln b - 6 \ln c - \frac{1}{2} \ln d$

   8. $\log_b b = 1$ since $b^1 = b$. $\log_7 7 = 1$. $\ln e = 1$. $\log 10 = 1$.

   9. $\log_b 1 = 0$ since $b^0 = 1$. $\log_8 1 = \log 1 = \ln 1 = 0$.

10. Log is a one-to-one function. This means if log c = log d, c = d.

NOTE

Not everything is one-to-one. If $x^2 = y^2$, $x = \pm y$.

11. Log is an increasing function. If m < n, then log m < log n.

12. $\lim\limits_{x \to \infty}(1 + 1/x)^x = e$

13. $b^{\log_b x}$ is a weird way of writing x. $e^{\ln x} = x$.

14. $\log_b b^x = x$; $\ln e^x = x$.

15. $\log_d c = \dfrac{\log_b c}{\log_b d}$; $\log_d c = \dfrac{\ln c}{\ln d}$; $\log_2 5 = \dfrac{\ln 5}{\ln 2}$.

You should now be able to solve the following kinds of log equations:

EXAMPLE 2—

Solve for x:

$4 \cdot 3^{x+2} = 28$

$3^{x+2} = 7$

$(x + 2) \log 3 = \log 7$

$x = \dfrac{\log 7 - 2\log 3}{\log 3}$

16. $a^x = e^{x \ln a}$. Also $x^x = e^{x \ln x}$ and $x^{\sin x} = e^{\sin x \ln x}$.

Divide by 4; isolate the exponent.

Take logs. It now becomes an elementary algebra equation, which we solve for x, using the same technique as in the implicit differentiation section of *Calc I*.

EXAMPLE 3—

$4^{2x+5} = 9^{3x-7}$

$(2x + 5) \log 4 = (3x - 7) \log 9$

Using the same algebraic tricks, we get

$$x = \dfrac{-7\log 9 - 5\log 4}{2\log 4 - 3\log 9} \quad \text{or} \quad \dfrac{7\log 9 + 5\log 4}{3\log 9 - 2\log 4}$$

Eliminate excess minus signs.

All this should be known about logs before the calculus. Now we are ready to get serious.

17. Major theorem: Given $f(x) = \ln x$; then $f'(x) = 1/x$. Proof (worth looking at):

Definition of derivative

$$\lim_{\Delta x \to 0} \frac{f(x + \Delta x) - f(x)}{\Delta x} = \frac{\ln(x + \Delta x) - \ln x}{\Delta x}$$

Rule 6

$$= \lim_{\Delta x \to 0} \frac{\ln\left(\dfrac{x + \Delta x}{x}\right)}{\Delta x}$$

Normal trick 1—multiply by 1 = x/x

$$= \lim_{\Delta x \to 0} \frac{x\ln\left(\dfrac{x + \Delta x}{x}\right)}{x\,\Delta x}$$

Algebra

$$= \lim_{\Delta x \to 0} \frac{\dfrac{x}{\Delta x}\ln\left(1 + \dfrac{\Delta x}{x}\right)}{x}$$

Rule 7

$$= \lim_{\Delta x \to 0} \frac{\ln\left(1 + \dfrac{\Delta x}{x}\right)^{x/\Delta x}}{x}$$

Use trick 2, $\dfrac{x}{\Delta x} = w$

As $\Delta x \to 0, w \to \infty$

$$= \lim_{w \to \infty} \frac{\ln\left(1 + \dfrac{1}{w}\right)^{w}}{x}$$

Rule 12

$$= \frac{\ln e}{x}$$

Rule 8

$$= \frac{1}{x}$$

This theorem is important, since it has a lot of log rules together with two normal math tricks. The theorem gives us the following result:

$$\int \frac{1}{x}\, dx = \ln|x| + C$$

CHAPTER 8

DERIVATIVES OF e^x, a^x, LOGS, TRIG FUNCTIONS, ETC.

We will now take derivatives involving ln x, e^x, a^x, $f(x)^{g(x)}$, trig functions, and inverse trig functions.

EXAMPLE 1—

$y = \ln(x^2 + 5x + 7)$

Let $u = x^2 + 5x + 7$. Then $y = \ln u$. So $dy/dx = (dy/du)(du/dx) = (1/u)(2x + 5) = (2x + 5)/(x^2 + 5x + 7)$.

Note that taking derivatives of logs is not difficult. However, you do not want to substitute $u = x^2 + 5x + 7$. You must do that ln your head. If $y = \ln u$, do $y' = (1/u)(du/dx)$ in your head!

EXAMPLE 2—

$$y = \ln \frac{(x^2 + 7)^9 (x + 3)}{x^6}$$

The simplest way to do this is to use laws 5, 6, and 7 of Chapter 7 and simplify the expression before we take the derivative. So $y = 9 \ln (x^2 + 7) + \ln (x + 3) - 6 \ln x$. Therefore

Remember to simplify by multiplying 9(2x) = 18x.

$$y' = \frac{9(2x)}{x^2 + 7} + \frac{1}{x + 3} - \frac{6}{x}$$

EXAMPLE 3

$y = \log_2 x$

Using law 15, $y = \ln x/\ln 2$, where $\ln 2$ is a number (a constant). Therefore

$$y' = (1/\ln 2)(\ln x)' = 1/(x \ln 2)$$

EXAMPLE 4

$F(x) = x \ln (\ln x)$

$$F'(x) = x\left(\frac{1}{\ln x} \cdot \frac{1}{x}\right) + \ln (\ln x)$$

$$F'(e) = e \cdot \frac{1}{\ln e} \cdot \frac{1}{e} + \ln (\ln e) = 1 + 0 = 1$$

since $\ln e = 1$ and $\ln 1 = 0$.

LAW 1

$y = e^u$. $y' = e^u (du/dx)$. If $y = e$ to the power u, where $u = $ a function of x, the derivative is the original function untouched times the derivative of the exponent.

EXAMPLE 5

$y = e^{\sin x}$

$y' = e^{\sin x}(\sin x)' = e^{\sin x}(\cos x)$

LAW 2

$y = a^u$. $y' = a^u \ln a (du/dx)$. If $y = a^u$, the derivative is a^u (the original function untouched) times the log of the base times the derivative of the exponent.

EXAMPLE 6—

$y = 7^{x^2}$

$y' = (7^{x^2})(\ln 7)(2x)$

   = (whole function) (ln of base) (derivative of exponent)

Let us, for completeness, recall the trig derivatives and do one longish chain rule.

   We are finally ready to finish a theorem we started near the beginning of the book.

THEOREM

If $y = x^n$, for n any real number, then $y' = nx^{n-1}$.

   Let $y = x^n$. Then $\ln y = n \ln x$. Differentiating implicitly, we get

$\dfrac{1}{y}\dfrac{dy}{dx} = \dfrac{n}{x}$. So $\dfrac{dy}{dx} = \dfrac{ny}{x}$ (New York over x). But $y = x^n$.

Substituting, we get $\dfrac{dy}{dx} = \dfrac{nx^n}{x} = nx^{n-1}$.

   It took all this time to finish because we needed logs. Logs really help a lot!

LAW 3

   A. $y = \sin x,\ y' = \cos x$

   B. $y = \cos x,\ y' = -\sin x$

   C. $y = \tan x,\ y' = \sec^2 x$

   D. $y = \cot x,\ y' = -\csc^2 x$

   E. $y = \sec x,\ y' = \tan x \sec x$

   F. $y = \csc x,\ y' = -\cot x \csc x$

EXAMPLE 7—

$y = \tan^6 (4x^2 + 3x + 7)$

Since this is a function of a function, we must use the extended chain rule.

Let $u = \tan(4x^2 + 3x + 7)$. $y = u^6$ and $dy/du = 6u^5$. Let $v = 4x^2 + 3x + 7$. $u = \tan v$. $du/dv = \sec^2 v$ and $dv/dx = 8x + 3$. So

$$
\begin{aligned}
dy/dx &= (dy/du) \times (du/dv) \times (dv/dx) \\
&= 6u^5 \times \sec^2 v \times (8x + 3) \\
&= [6 \tan^5(4x^2 + 3x + 7)]\,[\sec^2(4x^2 + 3x + 7)]\,(8x + 3)
\end{aligned}
$$

Power rule—leave trig function and crazy angle untouched   Derivative of trig function— leave crazy angle untouched   Derivative of crazy angle

You should be able to do this without substituting for u and v. It really is not that difficult with a little practice.

LAW 4

A. $y = \sin^{-1} u$. $y' = \dfrac{1}{(1 - u^2)^{1/2}} \dfrac{du}{dx}$ for $|u| < 1$.

B. $y = \tan^{-1} u$. $y' = \dfrac{1}{1 + u^2} \dfrac{du}{dx}$.

C. $y = \sec^{-1} u$. $y' = \dfrac{1}{u(u^2 - 1)^{1/2}} \dfrac{du}{dx}$ for $|u| > 1$.

EXAMPLE 8—

$y = \tan^{-1}(e^{5x})$

$u = e^{5x} \qquad u' = e^{5x}\,5$

$y' = \dfrac{1u'}{1 + u^2} = \dfrac{5e^{5x}}{1 + (e^{5x})^2} = \dfrac{5e^{5x}}{1 + e^{10x}}$

EXAMPLE 9—

$y = (\tan^{-1} x)/(1 + x^2)$

Use the quotient rule.

$$y' = \frac{(1 + x^2)[1/(1 + x^2)] - (\tan^{-1}x)(2x)}{(1 + x^2)^2}$$

$$= \frac{1 - 2x\tan^{-1}x}{(1 + x^2)^2}$$

I like this one. I don't know why, but I do.

EXAMPLE 10—

$$y = \sin^{-1}(\cos^2 x)$$

$$u = \cos^2 x \qquad u' = 2(\cos x)(-\sin x)$$

$$y' = \frac{u'}{(1 - u^2)^{1/2}} = \frac{-2\cos x\sin x}{(1 - \cos^4 x)^{1/2}}$$

EXAMPLE 11—

$$y = \sec^{-1}(x^2 + 1)$$

$$u = x^2 + 1 \qquad u' = 2x$$

$$y' = \frac{u'}{u(u^2 - 1)^{1/2}} = \frac{2x}{(x^2 + 1)[(x^2 + 1)^2 - 1]^{1/2}}$$

If $y = f(x)^{g(x)}$, we take logs of both sides and differentiate implicitly (if you've forgotten implicit differentiation, see my *Calc I*).

EXAMPLE 12—

$$y = x^{\sin x}$$

$$\ln y = \sin x \ln x$$

$$\frac{1}{y}\frac{dy}{dx} = \frac{\sin x}{x} + (\ln x)\cos x$$

$$\frac{dy}{dx} = y\left[\frac{\sin x}{x} + (\ln x)\cos x\right]$$

$$= x^{\sin x}\left[\frac{\sin x}{x} + (\ln x)\cos x\right]$$

EXAMPLE 12, ALTERNATIVE METHOD—

$y = x^{\sin x} = e^{\sin x \ln x}$

$y' = e^{\sin x \ln x} (\sin x \ln x)'$

$\quad = e^{\sin x \ln x} [\sin x/x + (\ln x) \cos x]$

$\quad = x^{\sin x} [\sin x/x + (\ln x) \cos x]$

EXAMPLE 13—

$$y = \frac{(x^3 + 1)^7 (\sin^6 x)(x^2 + 4)^9}{e^{7x} x^8}$$

To take derivatives without logs is long and leads to errors. However, taking logs first and differentiating implicitly make things much shorter and easier.

$\ln y = 7 \ln (x^3 + 1) + 6 \ln \sin x$
$\qquad + 9 \ln (x^2 + 4) - 7x - 8 \ln x$

$\dfrac{1}{y} \dfrac{dy}{dx} = \dfrac{7(3x^2)}{x^3 + 1} + \dfrac{6(\cos x)}{\sin x} + \dfrac{9(2x)}{x^2 + 4} - 7 - \dfrac{8}{x}$

$\qquad \dfrac{dy}{dx} = y \left(\dfrac{21x^2}{x^3 + 1} + 6 \cot x + \dfrac{18x}{x^2 + 4} - 7 - \dfrac{8}{x} \right)$

This looks neat. But remember what y really is!! This method is still pretty short!

CHAPTER 9

SHORTER INTEGRALS

In most schools, the largest part of the second semester of a three-term calculus sequence involves integrals. It usually covers more than 50 percent of this course. It is essential to learn these shorter ones as perfectly as possible so that Chapter 6 will not be overwhelming. Also, it is impossible to put every pertinent example in without making the book too long. The purpose of this book is to give you enough examples that you can do the rest by yourself. If you think an example should be added, write me.

RULE I

$$\int \frac{f'(x)\,dx}{f(x)} = \ln |f(x)| + C$$

One of the first new things we look for is that the numerator is the derivative of the denominator. This gives us an ln for an answer.

EXAMPLE 1—

Let u = 5x² − 7 and
du = 10x dx.

$$\int \frac{x\,dx}{5x^2 - 7}$$

$$= \frac{1}{10} \int \frac{10x\,dx}{5x^2 - 7} = \frac{1}{10} \int \frac{du}{u}$$

$$= \frac{1}{10}\ln|u| + C = \frac{1}{10}\ln|5x^2 - 7| + C$$

EXAMPLE 2—

u = 1 + sin x and
du = cos x dx

Exclude x = 3π/2 and so
on. Then sin x ⊃ −1, so
the absolute value is not
needed in the answer.

$$\int \frac{\cos x\,dx}{1 + \sin x}$$

$$= \int \frac{du}{u} = \ln|u| + C = \ln(1 + \sin x) + C$$

EXAMLE 3—

This one looks kind of
weird. Sometimes we just
have to try something.
Let u = x^{1/2} + 3. (Note
that u = x^{1/2} will also
work.) du = ¹/₂x^{-1/2} dx, so
dx = 2x^{1/2} du.

$$\int \frac{6\,dx}{x^{1/2}(x^{1/2} + 3)}$$

$$= \int \frac{6(2x^{1/2}\,du)}{x^{1/2}u} = \int \frac{12\,du}{u}$$

$$= 12\ln u + C = 12\ln(x^{1/2} + 3) + C$$

Let's try a definite integral.

EXAMPLE 4—

u = ln x; du = (1/x) dx

$$\int_{e}^{e^e} \frac{dx}{x \ln x}$$

$$= \int \frac{1}{u}\,du = \ln u = \ln(\ln x)\Big[_{e}^{e^e}$$

$$= \ln(\ln e^e) - \ln(\ln e) = \ln e - \ln 1 = 1 - 0 = 1$$

EXAMPLE 5—

$$\int \left(\frac{x^2 - 4x + 3}{x - 2} \right) dx$$

$$= \int \left(x - 2 + \frac{-1}{x - 2} \right) dx$$

$$= \frac{x^2}{2} - 2x - \ln|x - 2| + C$$

EXAMPLE 6—

$$\int \frac{2x\, dx}{(x^2 + 3)^2}$$

$$= \int \frac{du}{u^2} = \int u^{-2} du = \frac{-1}{u} + C = \frac{-1}{x^2 + 3} + C$$

We need to divide the long way since the degree of the top is greater than that of the bottom:

$$\begin{array}{r} x - 2 + -1/(x - 2) \\ x - 2\overline{)x^2 - 4x + 3} \\ \underline{x^2 - 2x} \\ -2x + 3 \\ \underline{-2x + 4} \\ -1 \end{array}$$

$u = x^2 + 3; du = 2x\, dx$. But be careful! This is not a logarithm!! The exponent on the bottom is 2!!! It must be a 1 to be a log!!!

TRIG INTEGRALS

RULE 2

A. $\int \sin x\, dx = -\cos x + C$

B. $\int \cos x\, dx = \sin x + C$

C. $\int \sec^2 x\, dx = \tan x + C$

D. $\int \tan x\, dx = -\ln|\cos x| + C$ or $\ln|\sec x| + C$

E. $\int \csc^2 x\, dx = -\cot x + C$

F. $\int \cot x\, dx = \ln|\sin x| + C$

G. $\int \tan x \sec x\, dx = \sec x + C$

H. $\int \sec x\, dx = \ln|\tan x + \sec x| + C$

I. $\int \csc x \cot x\, dx = -\csc x + C$

J. $\int \csc x\, dx = -\ln|\cot x + \csc x| + C$ or $\ln|\csc x - \cot x| + C$

You must know these integrals perfectly!!

EXAMPLE 7

u = 4x; dx = ¼ du.

$$\int \cot 4x\, dx$$

$$= \frac{1}{4}\int \cot u\, du = \frac{1}{4}\ln|\sin u| + C = \frac{1}{4}\ln|\sin (4x)| + C$$

NOTE
Whenever you have the integral of one of these trig functions and there is a constant multiplying the angle, you must, by sight, integrate this without letting u equal the angle. Otherwise, the integrals in Chapter 6 will take forever.

EXAMPLE 8

This is the crazy angle substitution:
u = crazy angle = 1 −3x³;
du = −9x² dx.

$$\int x^2 \cos (1 - 3x^3)\, dx$$

$$= -\frac{1}{9}\int (-9x^2)\cos (1 - 3x^3)\, dx$$

$$= -\frac{1}{9}\int \cos u\, du = -\frac{1}{9}\sin u + C$$

$$= -\frac{1}{9}\sin (1 - 3x^3) + C$$

EXAMPLE 9

u = tan 2x;
du = 2 sec² 2x dx.

$$\int \tan 2x \sec^2 2x\, dx$$

$$= \frac{1}{2}\int u\, du = \frac{1}{2}\frac{u^2}{2} + C = \frac{1}{4}\tan^2 2x + C$$

EXAMPLE 10—

$$\int \frac{1 + \sin x \, dx}{\cos^2 x}$$

$$= \int \left(\frac{1}{\cos^2 x} + \frac{\sin x}{\cos^2 x} \right) dx$$

$$= \int (\sec^2 x + \tan x \sec x) \, dx = \tan x + \sec x + C$$

This one requires splitting the integrand into two fractions and uses identities.

It's an easy one if (a *big* if) you know your identities and trig integrals.

EXPONENTIAL INTEGRALS

RULE 3

$$\int e^{bx} \, dx = \frac{e^{bx}}{b} + C. \text{ Know this perfectly by sight!}$$

EXAMPLE 11—

$$\int e^{5x} \, dx = \frac{e^{5x}}{5} + C$$

RULE 4

$$\int a^{bx} \, dx = \frac{a^{bx}}{b \ln a} + C. \text{ Know this one perfectly also!}$$

EXAMPLE 12—

$$\int 7^{5x} \, dx = \frac{7^{5x}}{5 \ln 7} + C$$

EXAMPLE 13—

$$\int \frac{e^{7/x}}{x^2} dx$$

$$= \int -\frac{e^u}{7} \, du = -\frac{e^u}{7} + C = -\frac{e^{7/x}}{7} + C$$

Crazy exponent substitutions: $u = 7/x$; $du = -7/x^2 \, dx$.

EXAMPLE 14—

Crazy exponent: u = ln x; du = (1/x) dx.

$$\int \frac{7^{\ln x}}{x} dx$$

$$= \int 7^u \, du = \frac{7^u}{\ln 7} + C = \frac{7^{\ln x}}{\ln 7} + C$$

EXAMPLE 15—

Crazy exponent (only real choice) plus trig identity: u = sin x; du = cos x dx = dx/sec x.

$$\int \frac{e^{\sin x} \, dx}{\sec x}$$

$$= \int e^u \, du = e^u + C = e^{\sin x} + C$$

EXAMPLE 16—

One of my favorites. This one looks exactly like the one in Example 16A but is really different: u = e^{4x} + 1; du = 4e^{4x} dx.

$$\int \frac{e^{4x} \, dx}{(e^{4x} + 1)^2}$$

$$= \frac{1}{4} \int \frac{4e^{4x} \, dx}{(e^{4x} + 1)^2} = \frac{1}{4} \int \frac{du}{u^2} = \frac{-1}{4u} + C = \frac{-1}{4(e^{4x} + 1)} + C$$

This is one of my favorite examples.

$$\int_e^{e^e} \frac{dx}{x \ln x} \cdot u = \ln x; \, du = \frac{1}{x} dx.$$

$$\int_e^{e^e} \frac{dx}{x \ln x} = \int \frac{1}{u} \, du = \ln u = \ln (\ln x) \Big|_e^{e^e} = \ln (\ln e^e) - \ln (\ln e) =$$

$$\ln (e \ln e) - \ln 1 = \ln e - 0 = 1 - 0 = 1.$$

INVERSE TRIG FUNCTIONS

This part is the last of the basic integrals that you must know by sight. In some schools, all six inverse trig functions must be known; in others, you learn three; and in still others, side as my school, you learn two. We will do three—arcsin, arctan, and arcsec.

RULE 5

$$\int \frac{dx}{(1 - x^2)^{1/2}} = \sin^{-1}x + C \qquad \int \frac{dx}{(a^2 - x^2)^{1/2}} = \sin^{-1}\left(\frac{x}{a}\right) + C$$

$$\int \frac{dx}{1 + x^2} = \tan^{-1}x + C \qquad \int \frac{dx}{a^2 + x^2} = \frac{1}{a} \tan^{-1}\left(\frac{x}{a}\right) + C$$

$$\int \frac{dx}{x(x^2 - 1)^{1/2}} = \sec^{-1}x + C \qquad \int \frac{dx}{x(x^2 - a^2)^{1/2}} = \frac{1}{a} \sec^{-1}\left(\frac{x}{a}\right) + C$$

Memorize these also!

EXAMPLE 16A (THE NEXT ONE I LEFT OUT)—

$$\int \frac{(e^{4x} + 1)^2}{e^{4x}} \, dx$$

This is different.

$$= \int \frac{e^{8x} + 2e^{4x} + 1}{e^{4x}} \, dx = \int e^{4x} + 2 + e^{-4x} \, dx$$

$$= \frac{e^{4x}}{4} + 2x - \frac{e^{-4x}}{4} + C$$

These integrals are not long, but you must study them because there are a lot of differences.

You will be able to identify these integrals by sight with practice. As for me, I know arcsin and arctan very well, because I've practiced them. However, in all the years I've taught, no one has ever required arcsec, so I have to struggle, since I need practice also. Practice is what is needed!

EXAMPLE 17—

You must see this is an arcsin. $u = e^{3x}$;
$u^2 = (e^{3x})^2 = e^{6x}$;
$du = 3e^{3x}\, dx$;
$du/3 = e^{3x}\, dx$.

$$\int \frac{e^{3x}\, dx}{(1 - e^{6x})^{1/2}}$$

$$= \int \frac{(1/3)\, du}{(1 - u^2)^{1/2}} = \frac{1}{3} \arcsin u + C = \frac{1}{3} \arcsin (e^{3x}) + C$$

EXAMPLE 18—

You must see this is an arctan. $u = x^6$;
$u^2 = (x^6) = x^{12}$;
$du = 6x^5\, dx$;
$a = 7^{1/2}$ since $a^2 = 7$.

$$\int \frac{x^5\, dx}{7 + x^{12}}$$

$$= \frac{1}{6} \int \frac{du}{7 + u^2} = \frac{1}{6 \cdot 7^{1/2}} \tan^{-1}\left(\frac{x^6}{7^{1/2}}\right) + C$$

EXAMPLE 19—

This is harder to tell. It is an arcsec with $u = x^2$, $du = 2x\, dx$. Multiply top and bottom by $2x$.
$a = 11^{1/2}$.

$$\int \frac{dx}{x(x^4 - 11)^{1/2}}$$

$$= \frac{1}{2} \int \frac{2x\, dx}{x^2(x^4 - 11)^{1/2}} = \frac{1}{2} \int \frac{du}{u(u^2 - 11)^{1/2}}$$

$$= \frac{1}{2}\left(\frac{1}{11^{1/2}}\right) \text{arcsec}\left(\frac{u}{11^{1/2}}\right) + C = \frac{\text{arcsec}\,(x^2/11^{1/2})}{2 \cdot 11^{1/2}} + C$$

You can all do it if you concentrate and practice a little.

Warning!! Beware! Danger! Now that you know these three, be careful of those that look similar but are not arctrig functions.

EXAMPLE 20—

$$\int \frac{x\,dx}{x^2 + 4}$$

$$= \frac{1}{2}\int du/u = \frac{1}{2}\ln u + C = \frac{1}{2}\ln(x^2 + 4) + C$$

This looks like an arctan, but a u substitution will give us a log. $u = x^2 + 4$; $du = 2x\,dx$; $du/2 = x\,dx$.

EXAMPLE 21—

$$\int \frac{x\,dx}{(1 - x^2)^{1/2}}$$

$$= -\frac{1}{2}\int u^{-1/2}\,du = -u^{1/2} + C = -(1 - x^2)^{1/2} + C$$

This is not an arcsin. $u = (1 - x^2)$; $du = -2x\,dx$.

CHAPTER 10

EXPONENTIAL GROWTH AND DECAY

In every book on calculus, there is a little on differential equations, which are equations with derivatives. Usually, one chapter is devoted to this topic, which is almost never used. Parts of one or two other chapters may have differential equations in them. This topic is almost universally covered by all courses.

EXAMPLE 1—

The rate of change of marlenium is proportional to the amount.

Ten pounds of marlenium become 90 pounds in 4 hours.

   A. Write the equation.

   B. How many pounds of marlenium will there be in 10 hours?

   C. When will there be 500 pounds of marlenium?

This is how we proceed:

Integrate.

A. The differential equation to solve is $dM/dt = kM$, where k is a constant. We solve this by separation of variables.

$$\frac{dM}{M} = k\,dt$$

$\ln M = kt + C$

We need a trick. Let $C = \ln M_o$.

$\ln M = kt + \ln M_o$.

where M_o = the amount of marlenium at t = 0.

$\ln M - \ln M_o = kt$

By law 6 of logs,

$$\ln \frac{M}{M_o} = kt$$

By the definition of logs,

$$\frac{M}{M_o} = e^{kt} \qquad \text{or} \qquad M = M_o e^{kt}$$

$M_o = 10 \qquad M = 90 \qquad t = 4 \qquad M = M_o\,e^{kt}$

Divide by 10.

$90 = 10e^{4k}$

Take lns.

$9 = e^{4k}$

$\ln 9 = 4k \ln e \qquad \text{or} \qquad k = \ln \frac{9}{4}$

But $e^{\ln 9} = 9$.

$M = 10e^{(\ln 9/4)t}$

Whew!!!!!

sooo $M = 10(9^{t/4})$

B. $t = 10$

$M = 10(9^{10/4}) = 10(9^{5/2}) = 10(3^5) = 2430$ pounds

C. $M = 500$ pounds. Solve for t. $500 = 10(9^{t/4})$

$50 = 9^{t/4}$

$\ln 50 = (t/4) \ln 9$. So $t = 4 \ln 50/\ln 9 = 7.12$ hours, by calculator.

There is a simpler way to get part A. If you notice the numbers, you will see that 9 comes from 90/10. Although the time, 4 hours, originally is in the numerator, after the derivation, the 4 turns, up in the bottom. Sooo, by observation

M = original times (future amount/original)$^{t/time}$

$M = 10(90/10)^{t/4}$ $M = 10(9^{t/4})$

Let's try another one. Suppose 76 exponentially decays to 31 in 5 days.

The equation is $N = 76(31/76)^{t/5}$. Simple, isn't it?

EXAMPLE 2—

Radioactive strontium 90 exponentially decays. Its half-life is 28 years. After an atomic attack, strontium 90 enters all higher life and is not safe until it decreases by a factor of 1000. How many years will it take strontium 90 to decay to safe levels after an atomic attack?

The equation, the short way, is $S = S_o(\frac{1}{2})^{t/28}$. The $\frac{1}{2}$ is for the half-life, or one-half the amount of radioactivity.

We can let $S_o = 1000$ and $S = 1$ for a reduction factor of 1000.

$$1 = 1000(\frac{1}{2})^{t/28}$$

$$0.001 = (0.5)^{t/28}$$

$\ln(0.001) = (t/28) \ln(0.5)$. $t = 28 \ln(0.001)/\ln(0.5) = 279$ years to be safe.

We must truly be careful not to unleash nuclear bombs!!

Interest is also an exponential function. Simple interest = principal times rate times time. If $t = 1$ year, $i = pr$, and the total amount $A = p + pr = p(1 + r)$. In other words, after 1 year, the principal is multiplied by $1 = pr$. After 2 years? $A = p(1 + r)^2$. After t years? $A = p(1 + r)^t$.

Suppose we have compounding interest twice a year, or one-half the interest rate $(r/2)$, but twice as many periods $(2t)$. $A = p(1 + r/2)^{2t}$. Compounded n times a year, the formula is $A = p(1 + r/n)^{nt}$.

Finally, if the interest is compounded continuously, $n \to \infty$, and $A = pe^{rt}$.

NOTE 1

If you use a bank with simple interest, go to another bank.

NOTE 2

For all intents and purposes, daily compounding is continuous compounding unless you have $10 billion.

NOTE 3

For continuous compounding formula verification, look at L'Hopital's rule.

Let's try an example.

EXAMPLE 3—

Suppose you have $100,000 invested at 10 percent.

A. How much would you have after 10 years, compounding yearly? Continuously?

B. When would you double your money, compounded yearly? Continuously?

The solution is

A. $p = 100,000$, $r = 0.10$, $n = 1$

$A = p(1 + r)^n = 100,000(1 + 0.1)^{10} = \$259,374.25$

$A = pe^{rt} = (100,000)e^{0.1(10)}\qquad = \$271,828.18$

That's a difference of $12,453.93. Now all we need is

the $100,000.

B. A = 200,000; p = 100,000; r = 0.10

$\quad$ A = p(1 + r)^t; 200,000 = 100,000 (1.1)^t;

$\quad$ t = $\dfrac{\ln 2}{\ln 1.1}$ = 7.27 years

$\quad$ A = pe^{rt}; 200,000 = 100,000 e$^{0.1t}$;

$\quad$ t = $\dfrac{\ln 2}{0.1}$ = 6.93 years

Continuous compounding does pay.

CHAPTER 11

WHAT YOU SHOULD KNOW FROM BEFORE TO DO THE NEXT

We have now come to the part of the book that requires you to work harder than perhaps at any other time in the entire calculus sequence. We are about to embark on learning new, long integration techniques. Since the product and quotient rules do not hold for integrals, we are forced to learn many techniques, most of which are long.

To make these integrals shorter, we list some crucial facts from previous chapters. If you have properly learned them, this chapter will be much easier.

1. The definition of the six trig functions

2. The values of the six trig functions for multiples of 30, 45, 60, and 90 degrees unless your instructor allows you to cheat and use calculators

3. The derivatives of the six trig functions

4. For the last time, the following identities:

   A. $\sin x \csc x = 1$
   B. $\cos x \sec x = 1$
   C. $\tan x \cot x = 1$

D. $\tan x = \sin x / \cos x$

E. $\cot x = \cos x / \sin x$

F. $\sin^2 x + \cos^2 x = 1$

G. $1 + \tan^2 x = \sec^2 x$

H. $1 + \cot^2 x = \csc^2 x$

I. $\sin 2x = 2 \sin x \cos x$

NOTE

It is of interest to note that you really don't need to know cos 2x, as we will see shortly.

J. $\sin^2 x = \dfrac{1 - \cos 2x}{2}$

K. $\cos^2 x = \dfrac{1 + \cos 2x}{2}$

5. The beginning integrals

A. Integral of x^n for $n \neq -1$

B. Multiplying out

C. Dividing out

D. u substitution in parentheses

E. u substitution for a crazy angle

F. u substitution for a crazy exponent

6. Trig integrals

A. $\int \sin ax \, dx = (1/a)(-\cos ax) + C$

B. $\int \cos ax \, dx = (1/a)(\sin ax) + C$

C. $\int \sec^2 ax \, dx = (1/a)(\tan ax) + C$

D. $\int \csc^2 ax \, dx = (1/a)(-\cot ax) + C$

E. $\int \tan ax \, \sec ax \, dx = (1/a) \sec ax + C$

F. $\int \cot ax \, \csc ax \, dx = -(1/a) \csc ax + C$

G. $\int \tan ax \, dx = -(1/a) \ln |\cos ax| + C$

or $(1/a) \ln |\sec ax| + C$

H. $\int \cot ax\ dx = (1/a)\ln\ |\sin ax|\ + C$

I. $\int \sec ax\ dx = (1/a)\ln\ |\sec ax + \tan ax|\ + C$

J. $\int \csc ax\ dx = -(1/a)\ln|\csc ax + \cot ax|\ + C$ or

$\qquad\qquad = (1/a)\ln\ |\csc ax - \cot ax|\ + C$

7. *Definition*: certain values involving multiples of 30, 45, 60, and 90 degrees, and derivatives of arcsin, arctan, and arcsec

8. Inverse trig integrals

   A. $\displaystyle\int \frac{dx}{(a^2 - x^2)^{1/2}} = \arcsin\left(\frac{x}{a}\right) + C$

   B. $\displaystyle\int \frac{dx}{a^2 + x^2} = \left(\frac{1}{a}\right)\arctan\left(\frac{x}{a}\right) + C$

   C. $\displaystyle\int \frac{dx}{x(x^2 - a^2)^{1/2}} = \left(\frac{1}{a}\right)\text{arcsec}\left(\frac{x}{a}\right) + C$

9. Other integrals you should know:

   A. $\displaystyle\int \frac{dx}{x} = \ln\ |x|\ + C$

   B. $\displaystyle\int \frac{f'(x)}{f(x)}\ dx = \ln\ |f(x)|\ + C$

   C. $\int \sinh ax\ dx = \left(\frac{1}{a}\right)\cosh ax + C$

   D. $\int \cosh ax\ dx = \left(\frac{1}{a}\right)\sinh ax + C$

   E. $\int e^{ax}\ dx = \left(\frac{1}{a}\right)e^{ax} + C$

   F. $\displaystyle\int b^{ax}\ dx = \frac{b^{ax}}{a\ln b} + C$

It is quite a list, but, as you will see, all are needed.

CHAPTER 12

LONGER INTEGRALS

The AP has all but eliminated this chapter from its requirements. I believe this is a serious mistake.

Not mastering these longer techniques is one reason why so few students major in math-related courses. If you are going on in math-related courses, make sure you master these techniques as well as others in a standard Calc II course, also found in my Calc II book.

Many high schools still include two of the techniques in their courses. These techniques are integration by parts and partial fractions. In addition, u substitutions are also included since some may be part of an AP or a standard high school course. Finally, I couldn't help showing you that the area of a circle really is πr^2.

Let's begin.

INTEGRATION BY PARTS

As you will see, there is very little theory in this chapter—only *hard* work.

Integration by parts comes from the product rule for differentials, which is the same as the product rule for derivatives.

Let u and v be functions of x.

$$d(uv) = u\,dv + v\,du \qquad \text{or} \qquad u\,dv = d(uv) - v\,du$$

Integrating, we get

$$\int u\,dv = \int d(uv) - \int v\,du \qquad \text{or}$$

$$\int u\,dv = uv - \int v\,du$$

What have we done? In the first integral, we have the function u and the differential of v. In the last integral, we have the differential of u and the function v. By reversing the roles of u and v, we hope to either have a very easy second integral or be able to proceed more easily to an answer.

EXAMPLE 1

$$\int xe^{3x}\,dx$$

If a polynomial multiplies e^{ax}, sin ax, and cos ax, we always let u = polynomial and dv = e^{ax} dx, sin ax dx, or cos ax dx. In this example,

$$u = x \qquad dv = e^{3x}\,dx \qquad v = \frac{e^{3x}}{3} \qquad du = 1\,dx$$

$$\quad u \quad\ dv \quad\ u\ \ v \quad\quad v\ \ du$$

$$\int xe^{3x}\,dx = x\,\frac{e^{3x}}{3} - \int \frac{e^{3x}}{3}\,dx$$

$$= \frac{xe^{3x}}{3} - \frac{e^{3x}}{3} + C$$

EXAMPLE 2—

$$\int x^4 \, e^{3x} \, dx$$

We must let u be a polynomial and $dv = e^{3x} \, dx$ four times!! However, if you observe the pattern, in time you may be able to do this in your head. Yes, I mean *you*. Signs alternate, polynomials get the derivative taken, a 3 is multiplied on the bottom each time, and e^{3x} multiplies each term.

The answer is

$$\left(\frac{x^4}{3} - \frac{4x^3}{9} + \frac{12x^2}{27} - \frac{24x}{81} + \frac{24}{243} \right) e^{3x} + C$$

NOTE

If we have $\int f(x) \cdot g(x) \, dx$ and $f(x)$ is a polynomial and $g(x)$ is e^{kx}, sin kx, or cos kx, then we let u be a polynomial and $dv = g(x)$, and we integrate by parts, of course.

Next we will consider integrating the arcsin, arctan, and ln. If you had never seen them before, you probably would never guess that all are done by integration by parts, since there appears to be only one function. However, mathematicians, being clever little devils, invented a second function so that all three of these integrals can be done rather easily.

EXAMPLE 3—

$$\int \sin^{-1} \, dx$$

Let $\quad u = \sin^{-1}]x \quad\quad du = \dfrac{dx}{(1 - x^2)^{1/2}}$

Let $\quad dv = 1 \, dx!! \quad\quad v = x$

$$\underset{u}{\int} \sin^{-1} x \, dx = \underset{u}{\int} \underset{dv}{\sin^{-1} x \, dx} = \underset{v}{x} \underset{u}{\sin^{-1} x} - \int \underset{v}{x} \underset{du}{\frac{dx}{(1 - x^2)^{1/2}}}$$

$$= x \sin^{-1} x + \int \frac{-x \, dx}{(1 - x^2)^{1/2}}$$

$$w = 1 - x^2; \, dw = -2x \, dx$$

$$= x \sin^{-1} x + \frac{1}{2} \int \frac{-2x \, dx}{(1 - x^2)^{1/2}}$$

$$= x \sin^{-1} x + \frac{1}{2} \int w^{-1/2} \, dw$$

$$= x \sin^{-1} x + \frac{1}{2} \frac{w^{1/2}}{\frac{1}{2}} + C$$

$$= x \sin^{-1} x + w^{1/2} + C$$

$$= x \sin^{-1} x + (1 - x^2)^{1/2} + C$$

NOTE I

If we have $\int f(x) \cdot g(x) \, dx$, $f(x) =$ a polynomial or is not there $(=1)$, and $g(x) = \ln x$ or $\sin^{-1} x$ or $\tan^{-1} x$ or $\sec^{-1} x$, then we let dv be a polynomial or 1 and $u = g(x)$. Integrate by parts.

NOTE 2

Although Example 3 is relatively short, some of these are verrrry long and use techniques we will learn later in this chapter.

We will now do a more complicated problem, $\int e^{5x} \cos 3x \, dx$. Based on what we did before, we can take either function as u and the rest as dv. It turns out both approaches will work. However, the problem is not quite so easy, as we will see. Being a glutton for punishment, I will show that the problem can be done in two ways.

EXAMPLE 4—

$$\int e^{5x} \cos 3x \, dx \quad u = e^{5x} \quad du = 5e^{5x} \, dx \quad dv = \cos 3x \, dx$$

$$v = \frac{\sin 3x}{3}$$

$$\phantom{\int e^{5x} \cos 3x \, dx}\quad u \qquad dv \qquad\quad u \quad v \qquad\quad v \qquad du$$

$$\int e^{5x} \cos 3x \, dx \; = \; \frac{e^{5x} \sin 3x}{3} - \int \frac{\sin 3x}{3} (5e^{5x}) \, dx$$

$$= \frac{e^{5x} \sin 3x}{3} - \frac{5}{3} \int e^{ex} \sin 3x \, dx$$

At this point you might say, "This doesn't do anything for us." You'd be right. Let's do it again. We let $U = e^{5x}$ because, if we reversed, we would wind up with the original integral and would have accomplished nothing. $dV = \sin 3x \, dx$. $V = (-\cos 3x)/3$. (Note that, in the third line, the product of three minus signs is a minus.)

$$\phantom{\int e^{5x} \cos 3x \, dx = \frac{e^{5x} \sin 3x}{3} - \frac{5}{3}\int}\quad U \qquad dV$$

$$\int e^{5x} \cos 3x \, dx = \frac{e^{5x} \sin 3x}{3} - \frac{5}{3} \int e^{5x} \sin 3x \, dx$$

$$\phantom{= \frac{e^{5x} \sin 3x}{3} - \frac{5}{3} e^{5x}}\quad U \qquad dV$$

$$= \frac{e^{5x} \sin 3x}{3} - \frac{5}{3} e^{5x} \frac{(-\cos 3x)}{3}$$

$$\phantom{-\frac{5}{3}\int 5e^{5x}}\quad dU \qquad V$$

$$-\frac{5}{3} \int 5e^{5x} \frac{\cos 3x}{3} \, dx$$

$$\int e^{5x} \cos 3x \, dx = \frac{e^{5x} \sin 3x}{3} + \frac{5e^{5x} \cos 3x}{9}$$

$$-\frac{25}{9} \int e^{5x} \cos 3x \, dx$$

It looks as if we will be going forever. However, notice that the original integral and the last integral are the same except for a constant. Call the original integral I (for integral, of course). The last line becomes

$$I = \frac{e^{5x} \sin 3x}{3} + \frac{5e^{5x} \cos 3x}{9} - \frac{25}{9}I$$

Now I = (9/9)I, so

$$\frac{34}{9}I = \frac{e^{5x} \sin 3x}{3} + \frac{5e^{5x} \cos 3x}{9}$$

and

$$I = \int e^{5x} \cos 3x \, dx = \frac{9}{34}\left(\frac{e^{5x} \sin 3x}{3} + \frac{5e^{5x} \cos 3x}{9}\right) + C$$

$$= \frac{3e^{5x} \sin 3x}{34} + \frac{5e^{5x} \cos 3x}{34} + C$$

NOTE

You do not have to multiply out the last step, but I wanted to show you that doing the problem in two ways gives the same answer. Also note that you do not have to do the problem in two ways, and I am a little crazy to try.

$$\int e^{5x} \cos 3x \, dx \qquad dv = e^{5x} \qquad v = \frac{e^{5x}}{5}$$

$$u = \cos 3x \qquad du = -3 \sin 3x \, dx$$

$$\overset{u}{} \quad \overset{dv}{} \qquad \overset{u}{} \quad \overset{v}{} \qquad \overset{v}{} \quad \overset{du}{}$$

$$\int \cos 3x \, e^{5x} \, dx = \frac{\cos 3x \, e^{5x}}{5} - \int \frac{e^{5x}}{5}(-3 \sin 3x) \, dx$$

$$\overset{dv}{} \quad \overset{u}{} \qquad\qquad\qquad \overset{dV}{} \quad \overset{U}{}$$

$$\int e^{5x} \cos 3x \, dx = \frac{\cos 3x \, e^{5x}}{5} + \frac{3}{5}\int e^{5x} \sin 3x \, dx$$

$$dV = e^{5x} \quad V = \frac{e^{5x}}{5} \quad U = \sin 3x \quad dU = 3\cos 3x \, dx$$

$$ V \quad U V dU$$

$$\int e^{5x}\cos 3x \, dx = \frac{e^{5x}\cos 3x}{5} + \frac{\tfrac{3}{5}\, e^{5x}\sin 3x}{5} - \frac{3}{5}\int \frac{e^{5x}\, 3\cos 3x \, dx}{5}$$

$$= \frac{e^{5x}\cos 3x}{5} + \frac{3e^{5x}\sin 3x}{25} - \frac{9}{25}\int e^{5x}\cos 3x \, dx$$

$$I = \frac{e^{5x}\cos 3x}{5} + \frac{3e^{5x}\sin 3x}{25} - \frac{9}{25}I \qquad I = \frac{25}{25}I$$

$$\frac{34}{25}I = \frac{e^{5x}\cos 3x}{5} + \frac{3e^{5x}\sin 3x}{25}$$

$$I = \int e^{5x}\cos 3x \, dx = \frac{25}{34}\left(\frac{e^{5x}\cos 3x}{5} + \frac{3e^{5x}\sin 3x}{25}\right) + C$$

$$= \frac{5e^{5x}\cos 3x}{34} + \frac{3e^{5x}\sin 3x}{34} + C$$

Our two answers check. Now that I've done it in two ways to show you that both ways give the same answer, I will never do it twice again!!!!!!!!

The last integration by parts, unless I think of another, is the integral of $\sec^3 x$. I think it more properly belongs to Example 7.

The next section involves integrals of trig functions. It is absolutely essential that you know the trig identities and integrals we listed before.

Let's consider integrals of the form $\sin^m x \cos^n x$.

EXAMPLE 5—

$\int \sin^6 x \cos^3 x \, dx$ ⠀⠀⠀when m or n is odd

The technique is to break off the trig function that has an odd power (if both are, break off the one that has the lower degree), and write all the others using the identity $\sin^2 x + \cos^2 x = 1$.

$$\int \sin^6 x \cos^3 x \, dx = \int (\sin^6 x)(1 - \sin^2 x)(\cos x) \, dx$$

$$u = \sin x \qquad du = \cos x \, dx$$

$$= \int u^6 (1 - u^2) \, du = \int (u^6 - u^8) \, du$$

$$= \frac{u^7}{7} - \frac{u^9}{9} + C$$

so

$$\int \sin^6 x \cos^3 x \, dx = \frac{\sin^7 x}{7} - \frac{\sin^9 x}{9} + C$$

Pretty simple, eh? However, when exponents m and n are even numbers, the integrals are usually much longer.

EXAMPLE 6—

$$\int \tan^3 x \sec^4 x \, dx$$

where m (power of tan) is odd and n (power of sec) is even. We can let $u = \tan x$ or $v = \sec x$. We will do it in both ways. Neither is too bad.

EXAMPLE 6A—

$$\int \tan^3 x \sec^4 x \, dx \qquad u = \tan x \qquad du = \sec^2 x \, dx$$

$$= \int (\tan^3 x)(\tan^2 x + 1) \sec^2 x \, dx$$

$$= \int u^3 (u^2 + 1) \, du = \int (u^5 + u^3) \, du = \frac{u^6}{6} + \frac{u^4}{4} + C$$

$$= \frac{\tan^6 x}{6} + \frac{\tan^4 x}{4} + C$$

EXAMPLE 6B—

$$\int \tan^3 x \sec^4 x \, dx \qquad v = \sec x \qquad dv = \tan x \sec x \, dx$$

$$= \int (\sec^2 x - 1)(\sec^3 x)(\tan x \sec x) \, dx$$

$$= \int (v^2 - 1)(v^3) \, dv = \int (v^5 - v^3) dv = \frac{v^6}{6} - \frac{v^4}{4} + C$$

$$= \frac{\sec^6 x}{6} - \frac{\sec^4 x}{4} + C$$

You might try to show that the answers to Examples 6A and 6B are the same, by using the identity $\sec^2 x = \tan^2 x + 1$.

EXAMPLE 7—

$$\int \sec^3 x \, dx$$

This is the worst case: m, the power of tan x, is even—m = 0—and n, the power of sec x, is odd—n = 3. All cases where m is even and n is odd are done by integrating by parts. They get long fast as the powers of m and n increase, and all involve the same tricks.

$$\overset{u}{} \qquad \overset{dv}{} \qquad \overset{u}{} \ \overset{v}{}$$

$$\int \sec^3 x \, dx = \int \sec x \sec^2 x \, dx = \sec x \tan x$$

$$\overset{v}{} \qquad \overset{du}{}$$

$$- \int (\tan x) \tan x \sec x \, dx$$

$$= \sec x \tan x - \int [(\sec^2 x - 1) \sec x] dx$$

$$= \sec x \tan x - \int \sec^3 x \, dx + \int \sec x \, dx$$

$$= \sec x \tan x - \int \sec^3 x \, dx + \ln |\sec x + \tan x|$$

Solving for the unknown integral, we get

$$\int \sec^3 x \, dx = \frac{\sec x \tan x + \ln|\sec x + \tan x|}{2} + C$$

EXAMPLE 8—

$$\int \tan^2 x \csc x \, dx$$

Anytime you have a mixed integral, that is, where the tan is not with the sec, you will have to use trig identities and usually tricks and sometimes long problems involving techniques that may not have been done here yet. The one I've given is a rather mild one.

$$\int \tan^2 x \csc x \, dx = \int \frac{\sin^2 x}{\cos^2 x} \frac{1}{\sin x} \, dx$$

$$u = \cos x \qquad du = -\sin x \, dx$$

$$= \int \frac{-du}{u^2} = \int -u^{-2} \, du$$

$$= \frac{1}{u} = \frac{1}{\cos x} + C = \sec x + C$$

PARTIAL FRACTIONS

We now do the section like the least. It is uninteresting, unimaginative, frequently overly long, and . . . necessary. Unless we have only linear factors, it is best to avoid this technique, if possible.

We wish to do the integrals by partial fractions. Suppose we have $R(x) = P(x)/Q(x)$, where the degree of $P(x)$ is less than the degree of $Q(x)$. We wish to break up $R(x)$ into simpler rational fractions; each piece is called a *partial fraction*. There will be one or more pieces for each linear factor x + a or quadratic factor $x^2 + b^2$ of Q(x). Here's how it looks in a particular case:

$$\frac{P(x)}{x^2(x-3)^3(x^2+4)^2} = \frac{A}{x} + \frac{B}{x^2} + \frac{C}{x-3} + \frac{D}{(x-3)^2} + \frac{E}{(x-3)^3}$$

$$\underbrace{\qquad\qquad}_{\text{2 from x}} \quad \underbrace{\qquad\qquad\qquad\qquad}_{\text{3 from }(x-3)}$$

$$+ \frac{Fx+G}{x^2+4} + \frac{Hx+I}{(x^2+4)^2}$$

$$\underbrace{\qquad\qquad\qquad\qquad}_{\text{2 from }(x^2+4)}$$

Notice that each linear factor gives pieces with constants on top, and each quadratic factor gives pieces with first-degree polynomials on top. The bottoms of the partial fractions are powers of the factors running from 1 to the power that occurs in Q(x). The constants A, B, C, D, E, F, G, H, and I have to be solved for, which I hope you never have to do. If you added all the fractions on the right, you would get the left fraction.

One more thing: Suppose $Ax^3 + Bx^2 + Cx + D = 4x^3 - 7x - 1$. Two polynomials are equal if their coefficients match. So A = 4, B = 0, C = −7, and D = −1.

There are a number of techniques that will allow you to solve for A, B, C, and so on. Two of them (combinations of) will serve us best.

EXAMPLE 9

$$\int \frac{x^3 - 7x + 18}{x^2 - 9}\, dx$$

$$= \int \left(x + \frac{2x+18}{x^2-9} \right) dx$$

Since the degree of the top is greater than or equal to the degree of the bottom, long-divide the bottom into the top until the degree of the top is less than the degree of the bottom.

Look at the fractional part only.

We will solve for A and B in two different ways. We now add the fractions and equate the tops since the bottoms are the same.

$$\frac{2x + 18}{x^2 - 9} = \frac{A}{x - 3} + \frac{B}{x + 3}$$

$$2x + 18 = A(x + 3) + B(x - 3)$$

METHOD 1

Multiply out left side and group terms.

$$A(x + 3) + B(x - 3) = 2x + 18$$

Now match coefficients.

$$(A + B)x + (3A - 3B) = 2x + 18$$

Solve two equations in two unknowns. It is really important for your algebra to be good.

$$A + B = 2$$

$$3A - 3B = 18$$

$$3A + 3B = 6$$
$$3A - 3B = 18$$
$$6A = 24$$

Substitute in either equation.

$$A = 4$$

$$B = -2$$

METHOD 2

This is true for all values of x. If we substitute x = 3 in both sides and then x = −3 in both sides, we will get both A and B with no work.

$$A(x + 3) + B(x - 3) = 2x + 18$$

If $x = 3$, $A(3 + 3) + B(3 - 3) = 2(3) + 18$; $6A = 24$; $A = 4$.

If $x = -3$, $A(-3 + 3) + B(-3 - 3) = 2(-3) + 18$; $-6B = 12$;

$B = -2$.

This way is so much easier—why don't we always use it? It is only perfect if we have all linear factors to the first power. Otherwise, it will not totally work. If there are no linear factors, you can't use this method. That is why both methods are needed. Let us finally finish the problem!

$$\int \frac{(x^3 - 7x + 18)\,dx}{x^2 - 9} = \int x\,dx + \int \frac{4\,dx}{x - 3} + \int \frac{-2\,dx}{x + 3}$$

$$= \frac{x^2}{2} + 4\ln|x - 3| - 2\ln|x + 3| + C$$

Note how easy the calculus part is. The algebra can be overwhelming.

EXAMPLE 10—

$$\int \frac{(9x^2 - 5x + 19)\,dx}{(x^2 + 5)(x - 2)}$$

Notice the degree of the top (2) is less than the degree of the bottom (3), so long division is not necessary. The bottom is already factored. There is one quadratic factor and one linear factor. The form is

$$\frac{Ax + B}{x^2 + 5} + \frac{C}{x - 2}$$

$$\frac{9x^2 - 5x + 19}{(x^2 - 5)(x - 2)} = \frac{Ax + B}{x^2 + 5} + \frac{C}{x - 2}$$

$$= \frac{(Ax + B)(x - 2) + C(x^2 + 5)}{(x^2 + 5)(x - 2)}$$

We now multiply out the top on the right and set the left numerator equal to the right numerator.

$$9x^2 - 5x + 19 = (A + C)x^2 + (-2A + B)x + (-2B + 5C)$$

$$A + C = 9 \tag{1}$$

$$-2A + B = -5 \tag{2}$$

$$-2B + 5C = 19 \tag{3}$$

These three equations in three unknowns are not bad but not particularly nice to solve. So we can use the other technique. Going back to the original top on the right, we have $(Ax + B)(x - 2) + C(x^2 + 5) = 9x^2 - 5x + 19$. There is only one linear factor, $x - 2$, but it is enough. Substituting $x = 2$ in this equation, we get $[A(2) + B](2 - 2) + C(2^2 + 5) = 9(2)^2 - 5(2) + 19$. From this we get $9C = 45$, or $C = 5$. Substituting $C = 5$ into Equation (1), we get $A = 4$. Substituting $A = 4$ into Equation (2), we get $B = 3$.

$$\int \frac{9x^2 - 5x + 19}{(x^2 + 5)(x - 2)}\, dx = \int \left(\frac{4x + 3}{x^2 + 5} + \frac{5}{x - 2} \right) dx$$

Splitting the first fraction on the right gives

$$\int \frac{9x^2 - 5x + 19}{(x^2 + 5)(x - 2)}\, dx = 2\int \frac{2x\, dx}{x^2 + 5} + \int \frac{3\, dx}{x^2 + 5} + \int \frac{5\, dx}{x - 2}$$

$$= 2\ln(x^2 + 5) + \left(\frac{x}{5^{1/2}} \right) \arctan\left(\frac{x}{5^{1/2}} \right)$$

$$+ 5\ln|x - 2| + C$$

EXAMPLE 11—

$$\int \frac{(6x^2 + 17x - 5)\, dx}{(x - 3)(x + 2)^2}$$

We have two linear factors, and 1 is to the second power—soooo . . .

$$\frac{6x^2 + 17x - 5}{(x - 3)(x + 2)^2} = \frac{A}{x - 3} + \frac{B}{x + 2} + \frac{C}{(x + 2)^2}$$

$$= \frac{A(x + 2)^2 + B(x + 2)(x - 3) + C(x - 3)}{(x - 3)(x + 2)^2}$$

Multiply and group; we get

$$6x^2 + 17x - 5 = A(x + 2)^2 + B(x + 2)(x - 3) + C(x - 3) \quad (1)$$

$$= A(x^2 + 4x + 4) + B(x^2 - x - 6) + C(x - 3) \quad (2)$$

$$A + B = 6 \qquad\qquad (3)$$
$$4A - B + C = 17 \qquad\qquad (4)$$
$$4A - 6B - 3C = -5 \qquad\qquad (5)$$

Now there are two good numbers, -2 and 3, but, as we will see, 3 is enough. Putting $x = 3$ into both sides of Equation (1), we get $A = 4$. Putting $A = 4$ into Equation (3), we get $B = 2$. Putting $B = 2$ and $A = 4$ into Equation (4) or (5), we get $C = 3$.

$$\int \frac{(6x^2 + 17x - 5)\,dx}{(x + 2)^2(x - 3)} = \int \frac{4dx}{x - 3} + \int \frac{2\,dx}{x + 2} + \int \frac{3\,dx}{(x + 2)^2}$$

$$= 4\ln|x - 3| + 2\ln|x + 2| - \frac{3}{x + 2} + C$$

The last part of this long integral chapter is called Additional and Substitutions.

ADDITIONAL AND SUBSTITUTIONS

All the extra goes into this part, which makes it more miserable for you.

EXAMPLE 12—

$$\int x(x + 1)^{80}\,dx$$

Sometimes the simplest substitutions work. We let $u = x + 1$. $du = x$. This transfers the power to the monomial and allows us to multiply out the expression $(x = u - 1)$.

$$\int x(x + 1)^{80}\,dx = \int (u - 1)u^{80}\,du = \int (u^{81} - u^{80})\,du$$

$$= \frac{u^{82}}{82} - \frac{u^{81}}{81} + C$$

$$= \frac{(x + 1)^{82}}{82} - \frac{(x + 1)^{81}}{81} + C$$

Sharp-eyed readers will note that there are at least two other ways to do this problem. The first is to multiply out the original. This is dismissed on grounds of sanity. The second is to use parts.

EXAMPLE 13—

$$\int \frac{dx}{x^{1/2} - x^{1/3}}$$

This one is perhaps the easiest to identify. LCD for $\frac{1}{2}$ and $\frac{1}{3}$ is 6. So, we let $u = x^{1/6}$; $u^2 = (x^{1/6})^2 = x^{1/3}$; $u^3 = (x^{1/6})^3 = x^{1/2}$; and $u^6 = x$. So $6u^5\, du = dx$. Substituting, we get

$$\int \frac{dx}{x^{1/2} - x^{1/3}} = \int \frac{6u^5\, du}{u^3 - u^2} = 6 \int \frac{u^5\, du}{u^2(u - 1)}$$

$$= 6 \int \frac{u^3\, du}{u - 1}$$

$$\begin{array}{r} u^2 + u + 1 + 1/(u - 1) \\ u - 1 \overline{)\, u^3 } \\ \underline{u^3 - u^2} \\ u^2 \\ \underline{u^2 - u} \\ u \\ \underline{u - 1} \\ 1 \end{array}$$

$$= 6 \int [u^2 + u + 1 + 1/(u - 1)]\, du$$

$$= 6(u^3/3 + u^2/2 + u + \ln|u - 1|) + C$$

$$= 2(x^{1/6})^3 + 3(x^{1/6})^2 + 6x^{1/6} + \ln |x^{1/6} - 1| + C$$

$$= 2x^{1/2} + 3x^{1/3} + 6^{1/6} + \ln |x^{1/6} - 1| + C$$

Note the nice pattern of the coefficients and the exponents of this answer. Well, *I* like it!

EXAMPLE 14—

$$\int \frac{x^3\, dx}{(x^2 + 4)^{1/2}}$$

This integral can be done in two new ways, both of which are useful.

METHOD A

Let u = the whole radical.

$u = (x^2 + 4)^{1/2}$; $u^2 = x^2 + 4$; $x^2 = u^2 - 4$; $2x\,dx = 2u\,du$; or $x\,dx = u\,du$. Substituting yields

$$\int \frac{x^3\,dx}{(x^2 + 4)^{1/2}} = \int \frac{x^2 x\,dx}{(x^2 + 4)^{1/2}} = \int \frac{(u^2 - 4)u\,du}{u}$$

$$= \int (u^2 - 4)\,du = \frac{u^3}{3} - 4u + C$$

$$= \frac{(x^2 + 4)^{3/2}}{3} - 4(x^2 + 4)^{1/2} + C$$

METHOD B

Let v = what's under the radical sign.

$v = x^2 + 4$; $x^2 = v - 4$; and $2x\,dx = dv$. Substituting gives

$$\int \frac{x^3\,dx}{(x^2 + 4)^{1/2}} = \frac{1}{2} \int \frac{x^2(2x)\,dx}{(x^2 + 4)^{1/2}} = \frac{1}{2} \int \frac{(v - 4)\,dv}{v^{1/2}}$$

$$= \frac{1}{2} \int (v^{1/2} - 4v^{-1/2})\,dv = \frac{1}{2} \left(\frac{2}{3}v^{3/2} - 8v^{1/2} \right) + C$$

$$= \frac{1}{3}(x^2 + 4)^{3/2} - 4(x^2 + 4)^{1/2} + C$$

The sharp-eyed reader will discover there are many, many other ways to do this problem. When a publisher becomes smart enough to publish this book, this problem will become a contest.

AREA OF A CIRCLE

We will finish by showing that the area of a circle really is πr^2. We will find one-quarter of the area of the circle $x^2 + y^2 = r^2$ and then multiply it by 4.

EXAMPLE 15—

$x = r \sin u; dx = r \cos u \, du.$

$$4\int_0^r (r^2 - x^2)^{1/2} \, dx$$

$x = r, r = r \sin u, 1 = \sin u,$
$u = \pi/2; x = 0, 0 = r \sin u,$
$0 = \sin u, u = 0.$

$$= 4\int_0^{\pi/2} (r^2 \cos u)(r \cos u) \, du$$

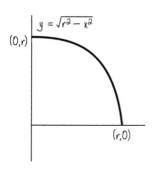

$$= 4r^2 \int_0^{\pi/2} \cos^2 u \, du$$

$$= 4r^2 \int_0^{\pi/2} \frac{(1 + \cos 2u) \, du}{2}$$

$$= 2r^2 \int_0^{\pi/2} (1 + \cos 2u) \, du$$

$$= 2r^2(u + \sin 2u)\Big[_0^{\pi/2} = 2r^2\left[\left(\frac{\pi}{2} + \sin \pi\right) - (0 - \sin 0)\right]$$

$$= 2r^2\left(\frac{\pi}{2}\right) = \pi r^2$$

The area of a circle really *is* πr^2, and you haven't been lied to all these years. It's nice to know.

CHAPTER 13

SECOND ODDS AND ENDS

L'HOPITAL'S RULE

Let us give a complete discussion of L'Hopital's rule since we now know logs.

L'HOPITAL'S RULE

1. $\displaystyle\lim_{x \to a} \frac{f(x)}{g(x)} = \frac{0}{0}$ or $\dfrac{\infty}{\infty}$

2. $\displaystyle\lim_{x \to a} \frac{f'(x)}{g'(x)} = L$

then $\displaystyle\lim_{x \to a} \frac{f(x)}{g(x)} = L$

The value a is any number or plus or minus infinity, and L could be any number or infinity.

This rule states that if the original limit is 0 over 0 or infinity over infinity or can be made into that form, then by taking the derivative of the top and the derivative of the bottom instead of using the quotient rule, we can find the limit of the original.

EXAMPLE 1—

$$\lim_{x \to 0} \frac{x}{x^2 + 1} = \frac{0}{1} = 0$$

No L'Hopital's rule.

EXAMPLE 2—

$$\lim_{x \to \infty} \frac{x^2 + 1}{1/x} = \frac{\infty}{0}$$

which means the limit does not exist. Also no L'Hopital's rule.

EXAMPLE 3—

$$\lim_{x \to 0} \frac{\sin x}{x} = \frac{0}{0}$$

Yay! L'Hopital's rule can be used.

$$\lim_{x \to 0} \frac{(\sin x)'}{x'} = \lim_{x \to 0} \frac{\cos x}{1} = \frac{1}{1} = 1$$

so $$\lim_{x \to 0} \frac{\sin x}{x} = 1$$

EXAMPLE 4—

Also 0/0.

$$\lim_{x \to 1} \frac{\ln x}{x - 1} \overset{\text{\textcircled{L}}}{=} \lim_{x \to 1} \frac{(\ln x)'}{(x - 1)'} = \lim_{x \to 1} \frac{1/x}{1} = 1$$

so $$\lim_{x \to 1} \frac{\ln x}{x - 1} = 1$$

EXAMPLE 5—

$$\frac{\infty}{\infty}$$

Here we need to use L'Hopital's rule twice.

$$\lim_{x\to\infty} \frac{2x^2 - 3x + 1}{5 - 7x^2} \overset{\text{\textcircled{L}}}{=} \lim_{x\to\infty} \frac{4x - 3}{-14\,x} \overset{\text{\textcircled{L}}}{=} \frac{4}{-14} = -\frac{2}{7}$$

so $\quad \lim_{x\to\infty} \dfrac{2x^2 - 3x + 1}{5 - 7x^2} = -\dfrac{2}{7}$

EXAMPLE 6—

$$\lim_{x\to 0}(x\ln x)$$

This turns out to be 0 times infinity (minus infinity). We must rearrange it so it is either 0/0 or ∞/∞. We use a little trick to make it ∞/∞.

$$\lim_{x\to 0} \frac{\ln x}{1/x} \overset{\text{\textcircled{L}}}{=} \lim_{x\to 0} \frac{(\ln x)'}{(1/x)'} = \lim_{x\to 0} \frac{1/x}{-1/x^2} = \lim_{x\to 0}(-x) = 0$$

so $\quad \lim_{x\to 0}(x\ln x) = 0$

EXAMPLE 7—

$$0 \times \infty; \lim_{x\to 0}(x\cot x) = \lim_{x\to 0} \frac{x}{\tan}$$

Trig identity. Now it's 0/0.

$$\lim_{x\to 0} \frac{x'}{(\tan x)'} = \frac{1}{\sec^2 x} = \frac{1}{1} = 1 \qquad \lim_{x\to 0}(x\cot x) = 1$$

EXAMPLE 8—

$$\lim_{x\to 0^+}\left(\frac{1}{\sin x} - \frac{1}{x}\right)$$

Trick—add the fractions, get 0/0.

$$\lim_{x\to 0^+} \frac{x - \sin x}{x\sin x} \overset{\text{\textcircled{L}}}{=} \lim_{x\to 0^+} \frac{1 - \cos x}{x\cos x + \sin x}$$

$$\overset{\text{\textcircled{L}}}{=} \lim_{x\to 0^+} \frac{\sin x}{2\cos x - (\sin x)x} = \frac{0}{2} = 0$$

Using L'Hopital's rule twice, we find the original limit to be 0.

If you exclude the log examples, all the preceding could have been done in the first semester. However, the following examples require logs. Those requiring logs are of the form 0^0, ∞^0, and 1^∞.

EXAMPLE 9—

0^0; $\lim\limits_{x \to 0^+} x^{4x}$; $y = x^{4x}$. $\ln y = 4x \ln x \to 0$ by Example 6. So,

$y \to e^0 = 1$.

EXAMPLE 10—

∞^0; $\lim\limits_{x \to 0^+} x^{1/x}$; $y = x^{1/x}$; $\ln y = (1/x) \ln x$; ∞/∞

$\lim\limits_{x \to \infty} \dfrac{(\ln x)'}{x'} = \lim\limits_{x \to \infty} \dfrac{1/x}{1} = \dfrac{0}{1} = 0$

Since $\ln y \to 0$, $y \to e^0 = 1$.

EXAMPLE 11—

1^∞; $\lim\limits_{x \to \infty} [1 + (1/x)]^x$

$y = \left(1 + \dfrac{1}{x}\right)^x$; $\ln y = x \ln\left(1 + \dfrac{1}{x}\right) = \dfrac{\ln(1 + 1/x)}{1/x} = \dfrac{0}{0}$

Taking derivatives top and bottom, we get

$\dfrac{\dfrac{-1/x^2}{1 + 1/x}}{-1/x^2} = \dfrac{1}{1 + 1/x} \to 1$

Thus, $\ln y \to 1$, so $y \to e^1 = e$.

NOTE

$\lim\limits_{x \to \infty} \left(1 + \dfrac{a}{bx}\right)^{cx} = e^{ac/b}$

EXAMPLE 12—

0^∞. No L'Hopital's rule, because $0^k = 0$ for all positive k.

NOTE

We use L'Hopital's rule if we have $0/0$, ∞/∞, $0 \cdot \infty$, $\infty - \infty$, 0^0, ∞^0, or 1^∞. It's a no-no for $0/\infty$, $\infty/0$, $\infty \cdot \infty$, 0^∞, ∞^∞.

IMPROPER INTEGRALS

In discussing an improper integral, it would seem to be a good idea to recall what a proper integral is. In *Calc I,* we defined the integral of f(x) from a to b in this way: Break up the interval (a, b) into n parts. Let w_i be any point in the interval Δx_i. Form the sum $f(w_1)\Delta x_1 + f(w_2)\Delta x_2 + f(w_3)\Delta x_3 + \cdots + f(w_n)\Delta x_n$. Form the sum $\sum_{i=1}^{n} f(w_i)\Delta x_i$. If the limit exists as n goes to infinity and all the deltas go to zero, we have

$$\lim_{\substack{n \to \infty \\ \text{all } \Delta x's \to 0}} \sum_{i=1}^{n} f(w_i)\Delta x_i = \int_a^b f(x)\,dx$$

At the start, we usually take f(x) to be continuous, although that can be weakened. However, implied in the definition is that everything is *finite*; that is, both a and b are finite and f(x) is always finite. What happens if we have an infinity? In effect we close our eyes and pretend the infinity is not there. We then take the limit as we go to that infinity. If the limit gives us a single finite number, we will say the integral *converges* to that number. Otherwise, the integral *diverges.* Let us be more formal.

EXAMPLE 13—

$$\int_{-1}^{\infty} \frac{1\,dx}{1 + x^2}$$

We rewrite this as

$$\lim_{x \to \infty} \int_{-1}^{a} \frac{1\,dx}{1 + x^2} = \lim_{a \to \infty} \tan^{-1} x \Big|_{-1}^{a} = \lim_{a \to a} [\tan^{-1} a - \tan^{-1}(-1)]$$

$$= \frac{\pi}{2} - \left(\frac{-\pi}{4}\right) = \frac{3\pi}{4}$$

You might ask, "Are they all this easy?" In most books, the vast majority of improper integrals are relatively easy, to make sure that you understand what an improper integral is without worrying about a complicated integral.

In summary, this integral *converges* to the value $3\pi/4$.

EXAMPLE 14—

$$\int_{4}^{\infty} \frac{1}{x^{1/2}}\, dx$$

We write

$$\lim_{a \to \infty} \int_{4}^{a} \frac{1}{x^{1/2}}\, dx = \lim_{a \to \infty} \int_{4}^{a} x^{-1/2}\, dx$$

$$= \lim_{a \to \infty} 2x^{1/2} \Big|_{4}^{a} = \lim_{a \to \infty} 2a^{1/2} - 2(4)^{1/2}$$

But $a^{1/2}$ goes to infinity as $a \to \infty$. Therefore, this integral *diverges*.

NOTE

In the kind of integral seen in Example 14, if the exponent in the denominator is less than or equal to 1, the integral diverges. If the exponent is greater than 1, the integral converges.

EXAMPLE 15—

$$\int_{0}^{\infty} \cos x\, dx = \lim_{a \to \infty} \int_{0}^{a} \cos x\, dx = \lim_{a \to \infty} \sin x \Big|_{0}^{a}$$

$$= \lim_{a \to \infty} (\sin a - \sin 0)$$

This integral does not go to infinity. Yet it still diverges since, as a goes to infinity, sin a takes on every value between −1 and 1. The integral diverges because it doesn't go to one finite value.

If we have $\int_{-\infty}^{\infty} f(x)\,dx$, we break it up into two pieces, $\int_{-\infty}^{c} f(x)\,dx + \int_{c}^{\infty} f(x)\,dx$, where, for convenience, c is often 0 but certainly does not have to be.

EXAMPLE 16—

$$\int_{-\infty}^{\infty} e^{-x}\,dx$$

Whenever you have infinity at both ends, you should try to do the piece that diverges first. If you choose the piece that diverges, you do not have to do the other piece, since the integral diverges (the whole integral diverges). If the first piece converges, then you still must do the other piece. Sharp-eyed readers have probably spotted the fact that the negative infinity piece diverges since, roughly speaking, $e^{-(-\infty)}$ goes to infinity. Note $e^{-\infty}$ is 0.

Let us take a look at the other infinity kind of improper integral.

EXAMPLE 17—

$$\int_{1}^{5} \frac{1\,dx}{(x-1)^{1/2}}$$

The improper part, $f(x) = (x-1)^{-1/2}$, is infinite at x = 1.

$$= \lim_{a \to 1^{+}} \int_{a}^{5} (x-1)^{-1/2}\,dx = \lim_{a \to 1^{+}} 2(x-1)^{1/2}\Big|_{a}^{5}$$

$$= \lim_{a \to 1^{+}} [2(5-1)^{1/2} - 2](a-1)^{1/2}$$

$$= 4 - 0 = 4$$

This integral *converges* to 4.

NOTE

For this kind of improper integral, if the exponent in the denominator is less than 1, the integral converges; if the exponent is greater than or equal to 1, the integral diverges.

EXAMPLE 18—

First, note that this *is* an improper integral since f(3) is undefined. Second, most of the time, if one piece diverges, both diverge, so that it is not important which piece is chosen first.

$$\int_2^5 \frac{1\,dx}{(x-3)^3}$$

$$= \lim_{c \to 3^-} \int_2^c (x-3)^{-3}\,dx + \lim_{d \to 3^+} + \int_d^5 (x-3)^{-3}\,dx$$

$$= \lim_{c \to 3^-} -\frac{1}{2(x-3)^2}\bigg[_2^c - \lim_{d \to 3^+} \frac{1}{2(x-3)^2}\bigg[_d^5$$

$$= \lim_{c \to 3^-}\left[-\frac{1}{2(c-3)^2} + \frac{1}{2(2-3)^2} \right]$$

$$+ \lim_{d \to 3^+}\left[-\frac{1}{2(5-3)^2} + \frac{1}{2(d-3)^2} \right]$$

Each bracketed piece goes to infinity. The integral *diverges.* However, if you did the problem, you should calculate only one piece. Once it diverges, you need not do the other piece.

The next two examples are a couple of my most favorites. It makes you believe that mathematicians can do *anything.* This is not true, of course, but the examples are extraordinary.

EXAMPLE 19—

We will take an infinite area, rotate the region, and get a finite volume!!!!!!!!! Here's our infinite area:

$$\int_1^\infty \frac{1}{x}\,dx = \lim_{a\to\infty}\int_1^a \frac{1}{x}\,dx$$

$$= \lim_{a\to\infty} \ln x \Big|_1^a = \lim_{a\to\infty}(\ln a - \ln 1) \to \infty$$

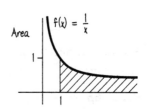

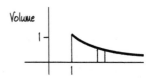

Volumes are determined by sections.

$$\pi\int_1^\infty \left(\frac{1}{x}\right)^2 dx = \lim_{a\to\infty}\pi\int_1^a x^{-2}\,dx = \lim_{a\to\infty}\pi\left[-\frac{1}{x}\right]_1^a$$

$$= \lim_{a\to\infty}\pi\left(-\frac{1}{a} + \frac{1}{1}\right) = \pi$$

Amazing!!!!!! Infinite area rotated gives finite volume.

This one will totally blow your mind. We will now take a finite region, rotate it, and get an infinite volume, which would seem impossible after the last example. It is *not*!!!!

EXAMPLE 20—

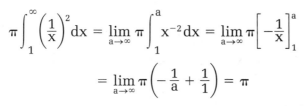

$$\int_0^1 \frac{1}{x^{2/3}}\,dx = \lim_{a\to 0^+}\int_a^1 x^{-2/3}\,dx = \lim_{a\to 0^+} 3x^{1/3}\Big|_a^1$$

$$= \lim_{a\to 0^+}(3 - 3a^{1/3}) = 3$$

Integral converges. Area is 3.

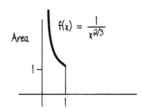

$$\pi\int_0^1 \left(\frac{1}{x^{2/3}}\right)^2 dx = \lim_{b\to 0^+}\pi\int_b^1 x^{-4/3}\,dx = \lim_{b\to 0^+}\pi\left(-\frac{3}{x^{1/3}}\right)\Big|_b^1$$

$$= \lim_{b\to 0^+}\pi\left(-3 + \frac{3}{b^{1/3}}\right) = \infty$$

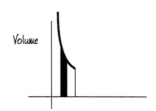

Diverges!!!! Amazing!!!!!

Here is one more example I like a lot because it is slightly nasty, slightly tricky, and verrry instructive.

EXAMPLE 21

Factor—improper part is at "infinity."

Break up by using partial fractions!

$$\int_0^\infty \frac{1}{x^2 + 3x + 2}\, dx$$

$$= \lim_{a\to\infty} \int_0^a \frac{1}{(x + 1)(x + 2)}\, dx$$

$$= \lim_{a\to\infty} \int_0^a \left(\frac{1}{(x + 1)} \frac{1}{(x + 2)} \right) dx$$

$$= \lim_{a\to\infty}\ \ln(x + 1) - \ln(x + 2)\Big|_0^a$$

$$= \lim_{a\to\infty}\ \ln(a + 1) - \ln(a + 2) - (\ln 1 - \ln 2)$$

Law 6 of logs!!!

Now $\lim\limits_{a\to\infty} \dfrac{a + 1}{a + 2} = 1.$

$$= \lim_{a\to\infty} \ln \frac{a + 1}{a + 2} + \ln 2$$

$$= \ln 1 + \ln 2 = \ln 2$$

Nice example!!!

SLOPE FIELDS

When the AP eliminated the long integrals, they needed a topic. So they added this one.

Most calculus books talk about vector fields. It is found at the end of many Calc III courses.

Let's get started.

Suppose we have been given the differential equation (an equation with a derivative in it)

$$\frac{dy}{dx} = \frac{x}{y^2 + 1}$$

Since dy/dx is the slope, substituting values (x, y) and drawing the slope at every point give you the slope field. Remember that the slope is the change in y over the change in x.

1. At any point on the y axis, where x = 0, m = 0. Let's choose other points.

2. At any point on the x axis, where y = 0, the slope is the x coordinate. For example, at the point (3, 0), the slope is 3.

3. At (1, 1), m = $^1\!/_2$. At (2, 1), m = $^2\!/_2$ = 1. At (3, 1), m = $^3\!/_2$. At (−1, 1), m = − $^1\!/_2$.

You keep on picking points until you get a picture.

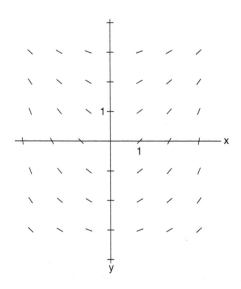

This particular curve can be solved exactly by a technique from differential equations (sometimes called Calc IV) called separation of variables. We've seen this a little before.

$$\frac{dy}{dx} = \frac{x}{y^2 + 1}$$

Cross-multiplying, we get x dx = (y² + 1) dy. So
$\int (y^2 + 1)\, dy = \int x\, dx$. So

$$\frac{y^3}{3} + y = \frac{x^2}{2} + C$$

Drawing the slope field gives an easier way to get a picture of the function than graphing the function, at least without a calculator.

The AP doesn't want you to do that much work. AP will probably give you a picture and ask you question(s) about it.

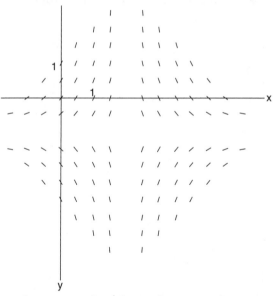

The AP might ask, "Where are the asymptotes?" If you look at the picture, the slopes go toward vertical at x = 2 and horizontal at y = −1, the two asymptotes.

EXAMPLE 22—

A. Graph y = xe⁻ˣ.

B. Find the area x ⩾ 0.

C. Find the volume if the region is rotated about the x axis.

Graphing exponentials is new, but we'll review graphing in general, areas, volumes of revolution, improper integrals, L'Hopital's rule, and integration by parts.

A. Graph $y = xe^{-x} = x/e^x$; $y' = (1 - x)e^{-x} = (1 - x)/e^x$; $y'' = (x - 2)e^{-x} = (x - 2)/e^x$.

1. x intercept(s) $y = 0$; top of fraction $= 0$; $x = 0$; point $(0, 0)$.

2. $(0, 0)$ is also the y intercept.

3. No vertical asymptotes, since the bottom of y is never 0.

4. Horizontal asymptote. This is different.

$$\lim_{x \to \infty} \frac{x}{e^x} \overset{\text{Ⓛ}}{=} \frac{1}{e^x} = 0$$

$$\lim_{x \to \infty} xe^{-x} \to -\infty$$

Asymptote is y = 0, but x is negative and e^{-x} gets very big as x goes to $-\infty$.

We have a one-sided asymptote. Many exponentials have a one-sided asymptote.

5. Possible max, min $y' = 0$; top $= 0$; $1 - x = 0$; $x = 1$; $y = 1/e^1$; $(1, 1/e)$; $y''(1) = (1 - 2)/e$ is negative, so $(1, 1/e)$ is a max.

6. Possible inflection points $y' = 0$. Again the top $= 0$; $x - 2 = 0$; $x = 2$; $y = 2/e^2$. The point is $(2, 2/e^2)$. It is an inflection point, since $y''(2^+)$ and $y''(2^-)$ are different signs.

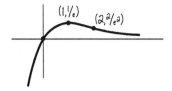

NOTE
Ⓛ, of course, means equal by L'Hopital's rule.

B. Find the area of $y = xe^{-x}$ where $x \geqslant 0$:

$$\int_0^\infty xe^{-x}\, dx$$

The improper part is infinity.

**Integrate by parts: u = x
and dv = e⁻ˣ.**

$$= \lim_{a \to \infty} \int_0^\infty xe^{-x}\,dx$$

$$= \lim_{a \to \infty}\left(\frac{xe^{-x}}{-1}\Big[_0^a + \int_0^a e^{-x}\,dx \right)$$

$$= \lim_{a \to \infty}\left(\frac{xe^{-x}}{-1}\Big[_0^a - e^{-x}\Big[_0^a \right)$$

$$= \lim_{a \to \infty}\left(-\frac{a}{e^a} + \frac{0}{e^0} - \frac{1}{e} + \frac{1}{e^0} \right)$$

$$= -0 + 0 - 0 + 1 = 1 \text{ square unit}$$

C. Find the volume if the region $y = xe^{-x}$ is rotated about the x axis, where $x \geqslant 0$.

**Improper part is again
infinity.**

**A ain, use integragation
by parts: u = x², dv = e⁻²ˣ.**

u = x, dv = e⁻²ˣ.

$$\pi \int_0^\infty r^2\,dx \qquad r = y = xe^{-x} \qquad r^2 = y^2 = x^2 e^{-2x}$$

$$= \lim_{a \to \infty} \pi \int_0^a x^2 e^{-2x}\,dx$$

$$= \lim_{a \to \infty}\left(\frac{\pi x^2 e^{-2x}}{-2}\Big[_0^a + \pi \int_0^a \frac{2xe^{-2x}}{2}\,dx \right)$$

$$= \lim_{a \to \infty}\left(\frac{\pi x^2 e^{-2x}}{-2}\Big[_0^a + \frac{\pi xe^{-2e}}{-2}\Big[_0^a + \pi \int_0^a \frac{e^{-2x}}{2}\,dx \right)$$

$$= \lim_{a \to \infty}\left(\frac{\pi}{-2}x^2 e^{-2x}\Big[_0^a - \frac{\pi}{2}xe^{-2x}\Big[_0^a - \frac{\pi}{4}e^{-2x}\Big[_0^a \right.$$

$$= \lim_{a \to \infty}\left[-\frac{\pi a^2}{2e^{4a}} + \frac{\pi 0^2}{2e^0} - \frac{\pi a}{2e^{4a}} + \frac{\pi(a)}{2e^0} - \frac{\pi(1)}{4e^{4a}} + \frac{\pi(1)}{4e^0} \right]$$

$$= -0 + 0 - 0 + 0 - 0 + \frac{\pi}{4}$$

The volume is $\pi/4$ cubic units.

CHAPTER 14

INFINITE SEQUENCES

This topic brings some controversy. Some people think it is very difficult. Others think it is very easy. I believe if you understand the beginning, the rest of the chapter is not too bad.

DEFINITION

Sequence—A sequence of terms, technically, is a function for which the domain is the positive integers. Nontechnically, there is a term called a_1 (read "$\bar{a}$ sub-1," where the "1" is a subscript not an exponent) denoting the first term, a_2 ("$\bar{a}$ sub-2") denoting the second term, and so on. The notation for an infinite sequence is $\{a_n\}$.

Let us give some examples. We will list some sequences, write the first four terms, and then write term number 100 by substituting 1, 2, 3, 4, . . . , 100 for n in a_n.

EXAMPLE 1—

$\{a_n\}$	1st	2nd	3rd	4th	100th
$\left\{\dfrac{n}{n+1}\right\}$	$\dfrac{1}{2}$	$\dfrac{2}{3}$	$\dfrac{3}{4}$	$\dfrac{4}{5}$	$\dfrac{100}{101}$
$\left\{\dfrac{(-1)^{n+1}(4n+1)}{n^2+1}\right\}$	$\dfrac{5}{2}$	$\dfrac{-9}{5}$	$\dfrac{13}{10}$	$\dfrac{-17}{17}$	$\dfrac{-401}{10{,}001}$
$\{6\}$	6	6	6	6	6

DEFINITION (NONTECHNICAL)

We write $\lim_{n\to\infty} a_n = L$ if as n gets larger, a_n gets closer to L.

In this case, we say that $\{a_n\}$ converges to L (or has the limit L). If a_n goes to plus or minus infinity or does not go to a single number, then $\{a_n\}$ diverges (or has *no* limit).

EXAMPLE 2—

Find the limit of $\{(n+9)/n^2\}$.

$a_n = (n+9)/n^2 = 1/n + 9/n^2$. As n goes to infinity, both terms go to 0. Therefore, the sequence converges to 0.

EXAMPLE 3—

Find the limit of $\{(2n^2 + 3n + 2)/(5 - 7n^2)\}$.

We divide top and bottom of a_n by n^2. We get $[2 + (3/n) + (2/n^2)]/(5/n^2 - 7)$. As n goes to infinity, a_n goes to $2/(-7)$. The sequence has the limit $-2/7$.

NOTE

This should look very familiar. This is how we found horizontal asymptotes. Also note that we can use L'Hopital's rule.

EXAMPLE 4

Find the limit of $\{a_n\} = \{\ln (n + 1) - \ln n\}$.

$$\lim_{n \to \infty} \ln (n + 1) - \ln n = \lim_{n \to \infty} \ln \left(\frac{n + 1}{n}\right)$$

$$= \lim_{n \to \infty} \ln \left(1 + \frac{1}{n}\right) = \ln 1 = 0$$

The sequence converges to 0.

EXAMPLE 5

Does $\{(-1)^n\}$ have a limit?

   This sequence is $-1, +1, -1, +1, -1, \ldots$. There is no limit because the sequence does not go to one number.

DEFINITION

$\lim_{n \to \infty} a_n = L$ if, given an $\varepsilon > 0$, there exists an $N > 0$ such that if $n > N$, $|a_n - L| < \varepsilon$.

NOTE

It is not important that you know the technical definition of a limit to understand the rest of the chapter. But . . . at this point of your mathematical career, you should start understanding the background. It probably will help you later on. It would also be nice if you could see the beauty and the depth of this material—the beginnings of calculus. It truly is a wonderful discovery.

EXAMPLE 6

Using ε, N, show $\lim_{n \to \infty} (2n + 5)/(n + 1) = 2$.

$$\left|\frac{2n + 5}{n + 1} - 2\right| = \left|\frac{2n + 5}{n + 1} - \frac{2(n + 1)}{n + 1}\right| = \left|\frac{3}{n + 1}\right| < \varepsilon$$

provided $3/\varepsilon < n + 1$ or $3/\varepsilon - 1 < n$. We then choose N as the whole-number part of $3/\varepsilon - 1$.

   The following theorems are used often. They are proved in many books and will only be stated here.

Let $\lim_{n\to\infty} a_n = L$, $\lim_{n\to\infty} b_n = M$; k = constant, f continuous. Then

1. $\lim_{n\to\infty} (a_n \pm b_n) = L \pm M$
2. $\lim_{n\to\infty} (a_n b_n) = LM$
3. $\lim_{n\to\infty} (a_n/b_n) = L/M \qquad M \neq 0$
4. $\lim_{n\to\infty} k a_n = kL$
5. $\lim_{n\to\infty} f(a_n) = f(L)$
6. $c_n \leq d_n \leq e_n$ and $\lim_{n\to\infty} c_n = \lim_{n\to\infty} e_n = P$. Then $\lim_{n\to\infty} d_n = P$.

EXAMPLE 7—

Show $\lim_{n\to\infty} (\sin n)/n = 0$.

Using part 6 above, $-1 \leq \sin n \leq 1$. So $-1/n \leq (\sin n)/n \leq 1/n$. As n goes to infinity, $-1/n$ and $1/n$ go to 0. Therefore, so does $(\sin n)/n$.

DEFINITION I

An *increasing sequence* is one where $a_n < a_{n+1}$ for all n.

DEFINITION 2

A *nondecreasing sequence* is one where $a_n \leq a_{n+1}$ for all n.

Similarly we can define *decreasing* and *nonincreasing*.

DEFINITION 3

A sequence is bounded if $|a_n| \leq M$ for some number M and all n.

Another theorem: Every bounded increasing (decreasing) sequence has a limit.

INFINITE SERIES, INCLUDING WHICH TEST TO USE (VERY IMPORTANT!)

I know this is getting to be a drag, but it is essential to understand the terminology. This understanding will make the rest of the chapter *much* easier. I don't know why, but it really seems to.

DEFINITION

Partial sums—Given sequence $\{a_n\}$:

1st partial sum $S_1 = a_1$

2nd partial sum $S_2 = a_1 + a_2$

3rd partial sum $S_3 = a_1 + a_2 + a_3$

nth partial sum $S_n = a_1 + a_2 + a_3 + \cdots + a_n = \sum_{k=1}^{n} a_k$

The *infinite series* $a_1 + a_2 + a_3 + \cdots +$ or $\sum_{k=1}^{\infty} a_k$ is said to *converge* to the sum S if $\lim_{n \to \infty} S_n = S$. If S does not exist, the series *diverges*.

EXAMPLE 8—

$0.767676 \cdots$

We can write this as an infinite series. $0.76 + 0.0076 + 0.000076 + \cdots$. This is a geometric series (infinite). This is one of the few series whose exact sum we can find.

$$S = \frac{a}{1 - r} \quad a = 0.76 \quad r = 0.01 \quad S = \frac{0.76}{1 - 0.01} = \frac{76}{99}$$

More generally, the series $a + ar + ar^2 + ar^3 + \cdots$ converges to $a/(1 - r)$ if $|r| < 1$.

EXAMPLE 9—

$4 - 8 + 16 - 32 + \cdots : a = 4, r = -2$; diverges.

EXAMPLE 10—

$1 + 1 + 1 + 1 + 1 + \cdots : a = 1, r = 1$; diverges.

EXAMPLE 11—

$1 - 1 + 1 - 1 + 1 - \cdots : a = 1, r = -1$; diverges.

Note that test 1 implies divergence in these three examples.

EXAMPLE 12—

$$\sum_{k=1}^{\infty} \frac{1}{k(k+1)}$$

Using partial fractions, we have

$$\frac{1}{k(k+1)} = \frac{1}{k} - \frac{1}{k+1}$$

Writing out the first few terms plus the $(n-1)$th term plus the nth term, we get

$$S_n = \left(\frac{1}{1} - \frac{1}{2}\right) + \left(\frac{1}{2} - \frac{1}{3}\right) + \left(\frac{1}{3} - \frac{1}{4}\right) + \cdots$$

$$+ \left(\frac{1}{n} - \frac{1}{n+1}\right)$$

Notice all the middle terms cancel out in pairs. So only the first and last terms remain:

$$S_n = 1 - \frac{1}{n+1} \qquad S = \lim_{n\to\infty} S_n = 1$$

Again, this is one of the few sequences whose exact value we can find. (This is called a *telescoping series*— it collapses like one of those toy or portable telescopes.) From this point on, for almost all the converging series, we will be able to tell that the series converges, but we won't be able to find its value. Later we will do some approximating.

EXAMPLE 13—

$$\sum_{k=1}^{\infty} \frac{4(2^k) + 5^k}{7^k}$$

After splitting, we get two geometric series:

$$S = \frac{4(2/7)}{1 - 2/7} + \frac{5/7}{1 - 5/7} = \frac{8}{5} + \frac{5}{2} = \frac{41}{10}$$

THEOREM

If $\sum_{n=1}^{\infty} a_n = L$ and $\sum_{n=1}^{\infty} b_n = M$, then $\sum_{n=1}^{\infty} (ca_n + b_n) = cL + M$

Now that we have an idea about what a sequence is and what an infinite series is (hopefully a *very good* idea), we would like to have some tests for when a series converges or diverges.

TEST 1

It is necessary that $a_k \to 0$ for $\sum_{k=1}^{\infty} a_k$ to converge.

NOTE 1

If a_k does not go to 0, $\sum_{k=1}^{\infty} a_k$ diverges.

NOTE 2

If a_k does go to 0 and that is all we know, we know nothing.

EXAMPLE 14—

Tell whether $\sum_{k=1}^{\infty} k/(k + 1)$ converges.

$k/(k + 1)$ goes to 1. Therefore $\sum_{k=1}^{\infty} k/(k + 1)$ diverges.

EXAMPLE 15—

The harmonic series $\sum_{k=1}^{\infty} (1/k)$.

Since $1/k$ goes to 0, we don't know if this series converges or diverges. We shall shortly show that the harmonic series diverges.

EXAMPLE 16—

The p_2 series $\sum_{k=1}^{\infty} (1/k^2)$.

Since $1/k^2$ goes to 0, again we can't tell. Shortly we shall show that the p_2 series converges.

TEST 2

Given a_k and $a_k > 0$, a_k goes to 0 for k big enough. Suppose we have a continuous function $f(x)$ such that $f(k) = a_k$. Then $\sum_{k=1}^{\infty} a_k$ and $\int_1^{\infty} f(x) \, dx$ either both converge or both diverge.

This theorem is easily explained by examples.

EXAMPLE 17—

Tell whether $\sum_{k=1}^{\infty} ke^{-k^2}$ converges or diverges.

The improper integral associated with $\sum_{k=1}^{\infty} ke^{-k^2}$ is $\int_1^{\infty} xe^{-x^2}\, dx$. Letting $u = -x^2$ and $du = -2x\, dx$, we find

$$\int_1^{\infty} xe^{-x^2}\, dx = \lim_{b \to \infty} \int_1^b \frac{-2xe^{-x^2}}{-2}\, dx = \lim_{b \to \infty} \left(-\frac{1}{2}\right) \int_{-1}^{-b^2} e^u\, du$$

$$= \lim_{b \to \infty}\left[-\frac{1}{2}(e^{-b^2} - e^{-1})\right] = \frac{1}{2e}$$

Since the improper integral converges, so does the infinite series.

NOTE I

The value of the improper integral is not the value of the infinite series. But we can say the following: If the integral and the series together converge, then $\int_1^{\infty} f(x)\, dx \le \sum_{k=1}^{\infty} a_k \le a_1 + \int_1^{\infty} f(x)\, dx$.

The bounds on $\sum_{k=1}^{\infty} ke^{-k^2}$ are

$$\frac{1}{2e} \le \sum_{k=1}^{\infty} ke^{-k^2} \le \frac{1}{e} + \frac{1}{2e}$$

NOTE 2

In this case, this is not too good an approximation.

We will get a better one if we take the fourth partial sum:

$$= \frac{1}{e} + \frac{2}{e^4} + \frac{3}{e^9} + \frac{4}{e^{16}}$$

The "error," the estimate on the rest of the terms, is

$$\sum_{k=5}^{\infty} ke^{-k^2} \le a_5 + \int_5^{\infty} xe^{-x^2}\, dx = \frac{5}{e^{25}} + \frac{1}{2e^{25}} \; 1.4 \times 10^{-11}$$

This is greater accuracy than you will probably ever need!!! Lots of things you cannot even integrate.

EXAMPLE 18—

The harmonic series $\sum_{k=1}^{\infty}(1/k)$ diverges.

$$\lim_{b\to\infty}\int_{1}^{b}\frac{1}{x}\,dx = \ln b$$

which goes to infinity as b goes to infinity.

EXAMPLE 19—

The p_2 series $\sum_{k=1}^{\infty}(1/k^2)$ converges.

$\int_{1}^{b}(1/x^2)\,dx = -1/b + 1$. Since $-1/b$ goes to 0 as b goes to infinity, this improper integral converges. So does the p_2.

EXAMPLE 20—

$$\sum_{k=1}^{\infty}\frac{1}{k^p}$$

If $p > 1$, it converges, and if $p \leq 1$, it diverges. Just use the integral test. It's easy.

TEST 3

The comparison test: Given $\sum_{k=1}^{\infty} a_k$, $\sum_{k=1}^{\infty} b_k$ where $0 < a_k \leq b_k$,

1. If $\sum_{k=1}^{\infty} b_k$ converges, so does $\sum_{k=1}^{\infty} a_k$.

2. If $\sum_{k=1}^{\infty} a_k$ diverges, so does $\sum_{k=1}^{\infty} b_k$.

Let us talk through part 1. The second part can be shown to be equivalent. The partial sums S_n of the $\sum_{k=1}^{\infty} b_k$ series are uniformly bounded because the first N terms are bounded by their maximum and the rest are bounded by $L + \varepsilon$. Therefore the partial sums of the $\sum_{k=1}^{\infty} a_k$ series also are bounded, being respectfully smaller than those of $\sum_{k=1}^{\infty} b_k$. Moreover, since $a_k > 0$, then the partial sums of the a_k form an increasing sequence. Now we have an increasing bounded sequence that has a limit. Therefore, $\sum_{k=1}^{\infty} a_k$ converges.

EXAMPLE 21

Examine $\sum_{k=1}^{\infty} 1/(4 + k^4)$.

$1/(4 + k^4) < 1/k^4$. $\sum_{k=1}^{\infty}(1/k^4)$ converges, by Example 20. Since the given series is smaller termwise than a convergent series, it must converge, by the comparison test.

EXAMPLE 22

Examine $\sum_{k=1}^{\infty}(2 + \ln k)/k$.

$\sum_{k=1}^{\infty}(2 + \ln k)/k > \sum_{k=1}^{\infty}(2/k)$ = twice a divergent series (the harmonic). Since the given series is larger termwise than a divergent series, the given series must diverge.

TEST 4

This is the limit comparison test. Given $\sum_{k=1}^{\infty} a_k$ and $\sum_{k=1}^{\infty} b_k$, $a_k \geq 0$, $b_k \geq 0$. If $\lim_{k \to \infty}(a_k/b_k) = r$, where r is any positive number, both series converge, or both diverge.

EXAMPLE 23

$$\sum_{k=1}^{\infty} \frac{3}{5k^4 + 4}$$

Let us compare this series with $\sum_{k=1}^{\infty}(1/k^4)$.

Divide top and bottom by k^4.

$$\lim_{k \to \infty}\left(\frac{3}{5k^4 + 4} \div \frac{1}{k^4}\right)$$

$$= \lim_{k \to \infty} \frac{3k^4}{5k^4 + 4}$$

$$= \lim_{k \to \infty} \frac{3}{5 + 4/k^4} = \frac{3}{5}$$

Since the limit is a positive number, both series do the same thing. Since $\sum_{k=1}^{\infty}(1/k^4)$ converges, so does $\sum_{k=1}^{\infty} 3/(5k^4 + 4)$.

TEST 5 (RATIO TEST)

Given $a_k \geq 0$, $\lim\limits_{k \to \infty} (a_{k+1}/a_k) = r$. If $r > 1$, it diverges.
If $r < 1$, it converges. If $r = 1$, use another test.

EXAMPLE 24—

Examine $\sum_{k=1}^{\infty} (k^2/5^k)$.

$$\frac{a_{k+1}}{a_k} = \frac{(k+1)^2}{5^{k+1}} \div \frac{k^2}{5^k}$$

$$= \frac{(k+1)^2}{5^{k+1}} \times \frac{5^k}{k^2} = \frac{k^2 + 2k + 1}{5k^2}$$

$$\lim_{k\to\infty} \frac{a_{k+1}}{a_k} = \lim_{k\to\infty} \frac{1 + \dfrac{2}{k} + \dfrac{1}{k^2}}{5} = \frac{1}{5} < 1 \qquad \sum_{k=1}^{\infty} \frac{k^2}{5^k}$$

The series converges.

EXAMPLE 25—

Examine $\sum_{k=1}^{\infty} (7^k/k!)$.

NOTE

6! Means 6(5)(4)(3)(2)(1).

ALSO NOTE:

$(k + 1)! = (k + 1)(k!)$. That is, $10! = 10(9!)$, and so on.

Let us again use the ratio test.

$$\frac{a_{k+1}}{a_k} = \frac{7^{k+1}}{(k+1)!} \div \frac{7^k}{k!}$$

$$= \frac{7^{k+1}}{(k+1)!} \times \frac{k!}{7^k} = \frac{7}{k+1}$$

$$\lim_{k\to\infty} \frac{7}{k+1} = 0 < 1 \qquad \sum_{k=1}^{\infty} \frac{7^k}{k!}$$

The series converges.

EXAMPLE 26—

Let's now look at $\sum_{k=1}^{\infty} (k^k/k!)$.

This is a little trickier than most. Again, we use the ratio test.

$$\frac{a_{k+1}}{a_k} = \frac{(k+1)^{k+1}}{(k+1)!} \div \frac{k^k}{k!}$$

$$= \frac{(k+1)^{k+1}}{(k+1)!} \times \frac{k!}{k^k} = \frac{(k+1)(k+1)^k k!}{(k+1)k! k^k} = \frac{(k+1)^k}{k^k}$$

$$= \left(1 + \frac{1}{k}\right)^k$$

Since the $\lim_{k \to \infty} [1 + (1/k)]^k = e > 1$, $\sum_{k=1}^{\infty} (k^k/k!)$ diverges.

EXAMPLE 27—

To show the third part of the previous theorem, you should apply the ratio test to both the harmonic series and the p_2 series. Both give a ratio of 1. The first series diverges, and the second converges. So, if the ratio is 1, we must indeed use another test.

TEST 6 (ROOT TEST)

Given $\sum_{k=1}^{\infty} a_k$. Take $\lim_{k \to \infty} (a_k)^{1/k} = r$ $(a_k \geq 0)$. If $r > 1$, it diverges. If $r < 1$, it converges. If $r = 1$, use another test.

NOTE

To show the third part $(r = 1)$, we would again use the harmonic and p_2 series. Let us give examples for the first two parts $(r > 1$ and $r < 1)$.

EXAMPLE 28—

$$\sum_{k=1}^{\infty} \frac{3^k}{k^k}$$

Take $(3^k/k^k)^{1/k} = 3/k$. $\lim_{k \to \infty} (3/k) = 0 < 1$. So the series converges.

EXAMPLE 29—

$$\sum_{k=1}^{\infty} \frac{2^k}{k^k}$$

Take $(2^k/k^2)^{1/k} = 2/k^{2/k}$. $\lim_{k\to\infty} (2/k^{2/k}) = 2/1 = 2$ $\left(\lim_{k\to\infty} k^{2/k} = \lim_{k\to\infty} e^{2(\ln k)/k} = e^{2(0)} = 1\right)$. Since $2 > 1$, the series diverges.

Up to this time, we have dealt exclusively with positive terms. Now we will deal with infinite series that have terms that alternate from positive to negative. We will assume the first term is positive. The notation will be as follows: alternating series $\sum_{k=1}^{\infty} (-1)^{k+1} a_k$, where all a_k are positive.

TEST 7

Given an alternating series where (1) $0 < a_{k+1} \le a_k$, $k = 1, 2, 3, 4, \ldots$, and (2) $\lim_{k\to\infty} a_k = 0$, the series converges to S and $S \le a_1$.

In clear English, the *only* thing you must do to show that an alternating series converges is to show the terms go to 0. (If only all series were that easy!)

EXAMPLE 30—

Alternating harmonic

$$\sum_{k=1}^{\infty} \frac{(-1)^{k+1}}{k}$$

converges since the terms go to 0.

DEFINITION

Absolutely convergent—A series $\sum_{k=1}^{\infty} a_k$ converges absolutely if $\sum_{k=1}^{\infty} |a_k|$ converges.

NOTE

If a series converges absolutely, it converges.

DEFINITION

Conditionally convergent—A series $\sum_{k=1}^{\infty} a_k$ converges conditionally if it converges but $\sum_{k=1}^{\infty} |a_k|$ diverges.

NOTE I

If we have an alternating series and want to show that it converges conditionally, we only have to show its terms go to 0. To find out whether it is absolutely convergent, we must use some other test.

NOTE 2

There are three possibilities for an alternating series: It diverges, converges conditionally, or converges absolutely.

Let us look at three alternating series.

EXAMPLE 31—

Let us look at $\sum_{k=1}^{\infty} [(-1)^{k+1}/(2k + 1)]$. This series converges conditionally since (1) the terms go to 0, but (2) by using the limit comparison test with the harmonic series, the positive series behaves as the harmonic series and diverges.

EXAMPLE 32—

What about the series $\sum_{k=1}^{\infty} [(-1)^{k+1}/(k^2 + 1)]$? This series converges absolutely by comparing to the p_2 series.

EXAMPLE 33—

$\sum_{k=1}^{\infty} [(-1)^{k+1}k^2/(k^2 + 6)]$ diverges since the terms don't go to 0.

DEFINITION

Region of convergence—We have an infinite series whose terms are functions of x. The set of all points x for which the series converges is called the *region of convergence*.

Now let's get back to the series with x's in them. Series of this type are usually done with the *ratio test*.

The purpose is to find the region of convergence. Then you will test both the left and right endpoints. There are three tests in all.

EXAMPLE 34—

$$\sum_{k=1}^{\infty} \frac{x^k}{k}$$

Using the ratio test,

$$\left| \frac{a_{k+1}}{a_k} \right| = \left| \frac{x^{k+1}}{k+1} \right| \div \left| \frac{x^k}{k} \right|$$

$$= \left| \frac{x^{k+1}}{k+1} \right| \times \left| \frac{k}{k^k} \right| = \left| \frac{kx}{k+1} \right| \cdot \lim_{k \to \infty} \left| \frac{kx}{k+1} \right| = |x|$$

So the region of convergence is $|x| < 1$ or $-1 < x < 1$.

Let us test both 1 and -1 by substituting those values into the original series. $x = 1$ gives us $\sum_{k=1}^{\infty}(1^k/k)$ or $\sum_{k=1}^{\infty}(1/k)$, the harmonic series that diverges. $x = -1$ gives us $\sum_{k=1}^{\infty}[(-1)^k/k]$, the alternating harmonic series, or rather the negative of the alternating harmonic series, since the first term is negative. We know this converges. Therefore, the region of convergence is $-1 \leq x < 1$.

IMPORTANT NOTE
When you test the endpoints, anything is possible. Both ends could converge, both could diverge, the left could converge and not the right, or the right could converge but not the left.

EXAMPLE 35—

Let's look at $\sum_{k=1}^{\infty}[(x-4)^k/3^k]$.

This is a geometric series that converges for

$$|r| = \left| \frac{x-4}{3} \right| < 1$$

Thus, the region of convergence is $|x - 4| < 3$ or $1 < x < 7$. Test $x = 1$ and substitute into the original series. We get

$$\sum_{k=1}^{\infty} \frac{(-3)^k}{3^k} \quad \text{or} \quad \sum_{k=1}^{\infty} (-1)^k = -1 + 1 - 1 + 1 - \cdots$$

which diverges (Example 11). For $x = 7$, we get

$$\sum_{k=1}^{\infty} \frac{3^k}{3^k} = 1 + 1 + 1 + 1 + \cdots$$

which diverges (Example 10).

EXAMPLE 36—

$\sum_{k=1}^{\infty} (x^k/k!)$, a nice one.

$$\left| \frac{a_{k+1}}{a_k} \right| = \left| \frac{x^{k+1}}{k + 1} \right| \div \left| \frac{x^k}{k!} \right|$$

$$= \left| \frac{x^{k+1}}{(k + 1)!} \right| \times \left| \frac{k!}{k^k} \right| = \left| \frac{xk!}{(k + 1)k!} \right| = \left| \frac{x}{k + 1} \right|$$

$$\lim_{k \to \infty} \left| \frac{x}{k + 1} \right| = 0$$

This says no matter what x is, the limit will always be less than 1. The region of convergence is *all real numbers.*

EXAMPLE 37—

$$\sum_{k=1}^{\infty} (k + 3)! \, x^k$$

$$\left| \frac{a_{k+1}}{a_k} \right| = |(k + 4)! \, x^{k+1}| \div |(k + 3)! \, x^k|$$

$$= \left| \frac{(k + 4)(k + 3)! \, x^{k+1}}{(k + 3)! \, x^k} \right| = |(k + 4)x|$$

$\lim_{k \to \infty} |(k + 4)x| \to \infty$ except if $x = 0$. The region of convergence is just the point $x = 0$.

Which Test to Use

After finishing the original draft of this book but before running off copies, I finished student-testing this section on infinite series. It became absolutely clear that this page is necessary.

1. First always see if the terms go to zero. If they don't, the series diverges. If the terms go to zero, the series at least converges conditionally if it alternates.

2. Use the integral test if the infinite series looks like an integral you have done. By this time, you should have so many integrals that you are familiar with and/or sick of them.

3. Don't use the integral test if you can see an easier one or if there is a factorial symbol.

4. My favorite is the ratio test. Always try the ratio test if there is a factorial or an x in the problem. Also try the ratio test if there is something to a power of k, such as 2^k or k^k.

5. Use the limit comparison test or the comparison test if the series looks like one you know such as the harmonic series, p_2 series, and so on. Use the comparison if the algebra is not too bad. Use the limit comparison if the algebra looks really terrible or even semiterrible.

6. Use the root test if there is at least one term with k in the exponent and no factorial in the problem.

7. If there is a series where there are a lot of messy-looking terms multiplying one another, the ratio test is probably the correct one.

8. Sometimes you may not be able to tell whether the terms go to zero. The ratio test may give absolute convergence or divergence immediately.

9. Practice the factorial. It is new to most of you. Once again, note that $(2n + 1)! = (2n + 1)(2n)! = (2n + 1)(2n)(2n - 1)!$, that is, $7! = 7(6!) = 7(6)(5!)$. Study factorial!!!!!

10. Most of all, do a lot of series testing. You will get better if you practice. The nice part is that the problems are mostly very short.

A PREVIEW OF POWER SERIES—TAYLOR'S THEOREM

We would like to have a polynomial approximation of a function in the vicinity of a given point. Polynomials are very easy to work with. They can be integrated easily, while many functions can't be integrated at all. Exact answers are usually not needed, since we do not live in a perfect world.

We therefore have Taylor's theorem, which give us a polynomial that approximates f(x) for every x approximately equal to a; the closer x is to a, the better the approximation for a given length. The "meat" of the theorem is a formula for the remainder, or error, when you replace the function by the polynomial. This is necessary so that you know how close your answer is.

Taylor's Theorem

1. $f^{(n+1)}(x)$, n + 1 derivatives, continuous on some interval I, where x is in the interval.

2. a is any number in interval I, usually its midpoint.

3. $S_n(x) = f(a) + \dfrac{f'(a)(x - a)}{1!} + \dfrac{f''(a)(x - a)^2}{2!}$

$+ \dfrac{f'''(a)(x - a)^3}{3!} + \cdots + \dfrac{f^{(n)}(a)(x - a)^n}{n!}$

NOTE

$S_n(x)$ is the sum of the polynomial terms up to the term of degree n.

4. The remainder $R_n(x) = f(x) - S_n(x)$ for all x in I.

Then there is a point w in I—w is between a and x— such that

$$R_n(x) = \frac{f^{(n+1)}(w)(x - a)^{n+1}}{(n + 1)!}$$

Let us give three examples worked out all the way.

EXAMPLE 38—

$f(x) = e^x \qquad a = 0$

Write a polynomial of degree 2. Write the remainder. Find the approximate value for $e^{0.2}$ and estimate the maximum error from the actual value of $e^{0.2}$.

This sounds like a lot of work, but as we will see, this process for e^x (e doesn't stand for "easy," but it should) is really quite short.

$f(x) = e^x \qquad f'(x) = e^x \qquad f''(x) = e^x \qquad f'''(x) = e^x$

$f(0) = f'(0) = f''(0) = 1 \qquad f'''(w) = e^w$

$$f(x) = f(a) + \underbrace{\frac{f'(a)(x - a)}{1!} + \frac{f''(a)(x - a)^2}{2!}}_{S_2(x)} + \underbrace{\frac{f'''(w)(x - a)^3}{3!}}_{R_2(x)}$$

$$e^x = 1 + \frac{x}{1} + \frac{x^2}{2} + \frac{e^w x^3}{6}$$

Therefore, $S_2(0.2) = 1 + 0.2 + (0.2)^2/2 = 1.22$ and $R_2(0.2) = e^w(0.2)^3/6$, where w is between 0 and 0.2. Because e^x is an increasing function, $e^w < e^{0.2} < e^{0.5} < (3)^{1/2} < 2$ (being wasteful). Therefore, $R_2(0.2) < 2(0.2)^3/6 \approx 0.00267$, the very maximum error.

This is a pretty good approximation. Remember, we were really rough-estimating the error, and this is only a polynomial of degree 2.

EXAMPLE 39—

Let us do the same for ln $(1 + x)$, polynomial of degree 3, $a = 0$, $x = 1$. Estimate the error for ln 1.1.

$$f(x) = \ln(x + 1) \quad f'(x) = (x + 1)^{-1} \quad f''(x) = (-1)(x + 1)^{-2}$$

$$f'''(x) = (-1)(-2)(x + 1)^{-3} \quad f''''(x) = (-1)(-2)(-3)(x + 1)^{-4}$$

$$f(0) = 0 \quad f'(0) = 1 \quad f''(0) = -1 \quad f'''(0) = 2$$

$$f''''(w) = -6(w + 1)^{-4}$$

$$f(x) = f(a) + f'(a)(x - a) + \frac{f''(a)(x - a)^2}{2!} + \frac{f'''(a)(x - a)^3}{3!}$$

$$+ \frac{f''''(w)(x - a)^4}{4!}$$

$$\ln(1 + x) = 0 + x - \frac{1x^2}{2} + \frac{2x^3}{6} - 6(w + 1)^{-4}\frac{x^4}{4!}$$

Therefore $S_3(x) = x - x^2/2 + x^3/3$ and ln $(1.1) = 0.1 - (0.1)^2/2 + (0.1)^3/3 = 0.098333 \cdots$. $R_3(x) = -x^4/4(w + 1)^4$.
    Let's estimate. $0 < w < 0.1$, so $1 < w + 1 < 1.1$. And sooooo, $1/1.1 < 1/(w + 1) < 1$; $|R_3(0.1)| = (0.1)^4/4$ $(w + 1)^4 < 0.1^4/4 = 0.000025$. Not bad!!

EXAMPLE 40—

Let's do the same for $f(x) = \sin x$ except let $a = 30° = \pi/6$ with a polynomial of degree 3, and $x = 32°$. Hold on to your hats, because this is pretty messy.

$$f(x) = \sin x \quad f'(x) = \cos x \quad f''(x) = -\sin x \quad f'''(x) = -\cos x$$

$$f^{(4)}(x) = \sin x \quad f\left(\frac{\pi}{6}\right) = \frac{1}{2} \quad f'\left(\frac{\pi}{6}\right) = \frac{3^{1/2}}{2} \quad f''\left(\frac{\pi}{6}\right) = \frac{1}{2}$$

$$f'''\left(\frac{\pi}{6}\right) = \frac{3^{1/2}}{2} \quad f^{(4)}(w) = \sin w$$

$$f(x) = f(a) + f'(a)(x - a) + \frac{f''(a)(x - a)^2}{2!} + \frac{f'''(a)(x - a)^3}{3!}$$

$$+ \frac{f^{(4)}(w)(x - a)^4}{4!}$$

$$\sin x = \frac{1}{2} + \frac{3^{1/2}}{2} \frac{(x - \pi/6)}{1!}$$

$$- \frac{\frac{1}{2}(x - \pi/6)^2}{2!} - \frac{(3^{1/2}/2)(x - \pi/6)^3}{3!}$$

$$+ \frac{(\sin w)(x - \pi/6)^4}{4!}$$

Now $x = 32° = 32\pi/180$. So

$$x - \frac{\pi}{6} = \frac{32\pi}{180} - \frac{30\pi}{180} = \frac{\pi}{90} = 0.035$$

Using the above approximation for $\sin x$, we get

$$\sin 32 = \frac{1}{2} + \left(\frac{3^{1/2}}{2}\right)(0.035) - \frac{\frac{1}{2}(0.035)^2}{2} - \frac{(3^{1/2}/2)(0.035)^3}{6}$$

$$= 0.5299985$$

The accuracy is not guaranteed, since $\pi/90$ should be to more places.

Actually, I couldn't bear this inaccuracy. So $\pi/90$ is 0.0349066 on my calculator, and if I hit the right buttons, $\sin 32° = 0.5299195$ (and my calculator is OK).

The remainder is $(\sin w) (0.0349066)^4/4!$. We know $\sin w$ is less than 1, so the remainder is less than $(0.0349066)^4/4!$ 6.1861288 $\times 10^{-8}$.

Even with these limited examples, we see different series get more accurate results with the same number of terms. This can be studied in great detail.

Let us try a short answer problem.

EXAMPLE 41—

$$f(x) = 1 + \frac{4(x - 4)}{1!} + \frac{7(x - 4)^2}{2!} + \frac{10(x - 4)^3}{3!}$$

$$+ \frac{13(x - 4)^4}{4!} + \cdots + \frac{(3n + 1)(x - 4)^n}{n!} + \cdots$$

What is $f'''(4)$?

When you take three derivatives, the third derivative of the first three terms is zero.

From the fifth term on, after three derivatives, there remains some power of x − 4.

Substituting x = 4, the derivatives of all remaining terms equal 0.

Let us concentrate on the one term of interest, $\dfrac{10(x-4)^3}{3!}$.

Taking three derivatives, we get $\dfrac{10(3)(2)(1)(x-4)^0}{3!} = 10$.

Also, it is very convenient to know certain power series by heart. We will list the most important together with the region of convergence.

e^x	All reals n 5 0, 1, 2, 3, . . .	$1 + x + \dfrac{x^2}{2!} + \dfrac{x^3}{3!} + \cdots + \dfrac{x^n}{n!} + \cdots$
$\sin x$	All reals k = 0, 1, 2, 3, · · ·	$\dfrac{x}{1!} - \dfrac{x^3}{3!} + \dfrac{x^5}{5!} - \dfrac{x^7}{7!} + \cdots + \dfrac{(-1)^k x^{2k+1}}{(2k+1)!} + \cdots$
$\cos x$	All reals k = 0, 1, 2, 3, · · ·	$1 - \dfrac{x^2}{2!} + \dfrac{x^4}{4!} - \dfrac{x^6}{6!} + \cdots + \dfrac{(-1)^k x^{2k}}{(2k)!} + \cdots$
$\ln(x+1)$	$-1 < x \le 1$ k = 1, 2, 3, 4, · · ·	$x - \dfrac{x^2}{2} + \dfrac{x^3}{3} - \dfrac{x^4}{4} + \cdots + \dfrac{(-1)^{k+1} x^k}{k} + \cdots$

Here are some more.

$\dfrac{1}{1-x}$	$-1 < x < 1$ k = 0, 1, 2, 3, · · ·	$1 + x + x^2 + x^3 + \cdots + x^k + \cdots$
$\dfrac{1}{x+1}$	$-1 < x < 1$	$1 - x + x^2\, x^3 + \cdots + (-1)^k x^k$ k = 0, 1, 2, 3, · · ·
Binomial $f(x) = (1+x)^p$	$-1 \le x < 1$	$1 + px + p(p-1)\dfrac{x^2}{2!}$ $+ \dfrac{p(p-1)(p-2)x^3}{3!} + \cdots$ $+ \dfrac{p(p-1)\cdots[p-(n-1)]x^n}{n!} + \cdots$ n = 0, 1, 2, 3, · · ·

Finally there are theorems found in many books that give the conditions under which you can add, subtract, multiply, divide, differentiate, and integrate infinite series. We can amaze ourselves by the number of functions we can approximate.

EXAMPLE 41—

Find the infinite series for cosh x.

$$\cosh x = \frac{(e^x + e^{-x})}{2}$$

$$e^x = 1 + x + \frac{x^2}{2!} + \frac{x^3}{3!} + \frac{x^4}{4!} + \cdots$$

$$e^{-x} = 1 - x + \frac{x^2}{2!} - \frac{x^3}{3!} + \frac{x^4}{4!} - \cdots$$

Substitute –x for x.

Add and divide by 2.

$$\cosh x = 1 + \frac{x^2}{2!} + \frac{x^4}{4!} + \frac{x^6}{6!} + \cdots$$

Pretty neat, eh?! More to come.

EXAMPLE 42—

$$\int_0^1 e^{-x^2}\, dx, \text{ four terms:}$$

$$e^x = 1 + x + \frac{x^2}{2!} + \frac{x^3}{3!}$$

$$e^{-x^2} = 1 - x^2 + \frac{x^4}{2} - \frac{x^6}{6}$$

$$\int_0^1 e^{-x^2}\, dx = x - \frac{x^3}{3} + \frac{x^5}{10} - \frac{x^7}{42}$$

$$= 1 - \frac{1}{3} + \frac{1}{10} - \frac{1}{42} = \frac{78}{105}$$

Not too shabby. More to come.

EXAMPLE 43—

The series for $x/(1 + x)^2$:

Differentiating, we get:

Multiply by –x; our result is:

$$\frac{1}{1 + x} = 1 - x + x^2 - x^3 + x^4 + \cdots$$

$$-\frac{1}{(1 + x)^2} = -1 + 2x - 3x^2 + 4x^3 - 5x^4 + \cdots$$

$$\frac{x}{(1 + x)^2} = x - 2x^2 + 3x^3 - 4x^4 + 5x^5 - \cdots$$

When mathematicians do things like this, you tend to believe that mathematics can do everything and anything. However, this is not true. However, the best is yet to come!!!!!!

We can derive every property of the sine and cosine by using infinite series, never, *never* mentioning triangles or angles. Amazing, huh?! Given $\sin x = x - x^3/3! + x^5/5! - \cdots$ and $\cos x = 1 - x^2/2! + x^4/4! - x^6/6! + \cdots$. How about $\cos 2x$? $\cos 2x = 1 - 2x^2 + (2/3)x^4 - (4/45)x^6 + \cdots$ [2x for x in cos x].

How about $\sin^2 x + \cos^2 x = 1$?

$$\sin^2 x + \cos^2 x = \left(x - \frac{x^3}{6} + \frac{x^5}{120} - \cdots\right)\left(x - \frac{x^3}{6} + \frac{x^5}{120} - \cdots\right)$$

$$+ \left(1 - \frac{x^2}{2} + \frac{x^4}{24} - \frac{x^6}{720} + \cdots\right)$$

$$\times \left(1 - \frac{x^2}{2} + \frac{x^4}{24} - \frac{x^6}{720} + \cdots\right)$$

$$= x^2 - \frac{x^4}{3} + \frac{2x^6}{45} - \cdots + 1 - x^2$$

$$+ \frac{x^4}{3} - \frac{2x^6}{45} + \cdots$$

$$= 1 + 0 + 0 + 0 + \cdots = 1$$

How about $\tan x$? Well we would like $\tan x = \sin x/\cos x$.

$$1 - x^2/2 + x^4/24 - x^6/72 + \cdots \overline{)x - x^3/6 + x^5/120 - x^7/540 \cdots}^{\displaystyle x + x^3/3 + 2x^5/15 + 17x^7/315 \cdots = \tan x}$$

How about derivatives? If $f(x) = \sin x$, we want $f'(x) = \cos x$.

$$\sin x = x - \frac{x^3}{6} + \frac{x^5}{120} - \frac{x^7}{5040} + \cdots$$

$$(\sin x)' = 1 - \frac{3x^2}{6} + \frac{5x^4}{120} - \frac{7x^6}{5040} + \cdots$$

$$= 1 - \frac{x^2}{2} + \frac{x^4}{24} - \frac{x^6}{720} + \cdots = \cos x$$

We could, of course, integrate $\cos x$ term by term and get $\sin x$. We could go on and on, getting every property of the sine and cosine and all other trig functions totally without angles or triangles. The beauty of these power series is that they are limits of polynomials and are easy to deal with. Yet there are things that are even more powerful in mathematics. But they must wait for another book.

INDEX